T0177796

PRAISE FOR *HUMAN FLOURISHING: SCIENTIFIC INSIGHT AND SPIRITUAL WISDOM IN UNCERTAIN TIMES*

'The struggle for human beings to integrate a thoughtful understanding of the world as described by science and an ambitious hope of human flourishing as described by philosophy or faith is one at which humans have largely failed over the last three hundred years. This book is a major step in the right direction. It is very serious about science and very serious about human beings and their hopes and fears. I warmly commend it for a careful and thoughtful provocation towards a deeper commitment to the flourishing of human beings and of the creation.'

Justin Welby, Archbishop of Canterbury

'The theme of this highly readable and enlightening book is broad and ambitious. It's the product of the authors' deep engagement with science, ethics and religion, and analyses the requisites for a fulfilled life, highlighting those that too often elude politicians and economists. The text is enlivened with historical allusions and quotations. It offers a wise perspective that's much needed as individuals and societies contend with the anxieties of the present era.'

Lord Martin Rees, Astronomer Royal,
former President of the Royal Society

'In this magisterial book, Andrew Briggs and Michael Reiss address one of the most fundamental issues confronting humanity—human flourishing. Drawing on science and religion, they examine it from the perspective of the material, relational and spiritual. What emerges are profound insights into meaning, purpose, truth, and the reason for being. This book should be read by anyone interested in what it is to be human.'

Colin Mayer, Peter Moores Professor of Management Studies,
University of Oxford

'What enables the good life? Material goods? Supportive relationships? Transcendent purpose? In this state-of-the-art synopsis, scientist Andrew Briggs and bioethicist Michael Reiss weave these and other threads into the fabric of human thriving. With a breath-taking sweep of scholarship that draws insights from multiple disciplines, they illuminate a path toward meaningful well-being and sustainable joy.'

David Myers, Professor of Psychology, Hope College,
author of *The Pursuit of Happiness*

'A sophisticated and much-needed and insightful integration of science and humanity. As an economist I am embarrassed by my profession's stunted characterization of humanity as "*Homo economicus*", which shrivels us to hedonistic consumers. In reality, as Professors Briggs and Reiss demonstrate, we thrive from morally guided agency that transcends ourselves and our time on Earth. In this time of uncertainty and pessimism, it is a hopeful guide to meaningful lives.'

Sir Paul Collier, Blavatnik School of Government,
author of *The Future of Capitalism*

'In a world where human flourishing seems somewhat more elusive and abstract than ever, Professors Briggs and Reiss capture the many dimensions of human flourishing in the 21st century. In doing so, they give us reason to hope and to work toward a world where all people flourish. This is a delightful and uplifting treatise on what it means to be human.'

Heather Templeton Dill, President, John Templeton Foundation

In *Human Flourishing: Scientific Insight and Spiritual Wisdom in Uncertain Times*, acclaimed scholars Andrew Briggs and Michael Reiss provide insight for navigating a world of uncertainty and complexity to find more meaning, purpose, and happiness all around us. Using a combination of science and ancient wisdom, they demonstrate why love is essential for human flourishing.

Arthur C. Brooks, Professor, Harvard Kennedy School and Harvard Business School, and *The New York Times* bestselling author

'For those of my generation, who grew up with post-war austerity and the threat of nuclear annihilation, the twenty-first century promised an era of unparalleled human flourishing. But it was a mirage. Material wealth has led to problems of disparity, over-consumption, and climate catastrophe; social media has produced alienation and a retreat from shared values. Democracy and common decency look increasingly fragile. We have entered a strange new era in which extraordinary promise is coupled with a burgeoning sense of insecurity and uncertainty. Science, the powerful facilitator of progress, also threatens our undoing. In this lucid and comprehensive analysis, Andrew Briggs and Michael Reiss carefully examine the rich tapestry of religious, cultural, and scientific factors that define our current predicament, and offer a message of hope, a way ahead founded on that familiar, yet too-often elusive, human quality—love.'

Paul Davies, Director of the Beyond Center for Fundamental Concepts in Science, Arizona State University

'This book by Briggs and Reiss covers questions that are of critical importance to everyone everywhere: How do we understand human life? What is human flourishing? How do we flourish? The book's rich insights and comprehensive scope will be of benefit to all readers. It provides a roadmap to flourishing in this life, and beyond.'

Tyler J. VanderWeele, Loeb Professor of Epidemiology and Director of the Human Flourishing Program, Harvard University

'In the midst of a great pandemic, unprecedented poverty, and natural disasters alongside never-before-seen development of new technologies and great wealth, nothing could be more important than wrestling with what it really means for humans to flourish. Here, Briggs and Reiss provide a comprehensive, synthetic, and highly readable book that addresses this topic head on. It is the kind of book that should be read and re-read.'

Elaine Howard Ecklund, Herbert S. Autrey Chair in Social Sciences, Rice University

'As I read this book, Modest Mussorgsky's wonderful *Pictures at an Exhibition* started playing in my mind. The same sense of multiple perspective, overt spaciousness with periodic attention to intense detail, yet a persistent crescendo in continuity of purpose emerges in this elegant and comprehensive tour of a rich and pan-disciplinary subject. Briggs and Reiss have given a compelling introduction to human flourishing, and show us why, though discussed since the ancient world, it has become ever more pressing in our own times.'

Tom McLeish, Professor of Natural Philosophy, University of York

HUMAN FLOURISHING

Scientific Insight and Spiritual Wisdom
in Uncertain Times

Andrew Briggs

Professor of Nanomaterials, University of Oxford

Michael J. Reiss

Professor of Science Education, University College London

OXFORD
UNIVERSITY PRESS

OXFORD
UNIVERSITY PRESS

Great Clarendon Street, Oxford, OX2 6DP,
United Kingdom

Oxford University Press is a department of the University of Oxford.
It furthers the University's objective of excellence in research, scholarship,
and education by publishing worldwide. Oxford is a registered trade mark of
Oxford University Press in the UK and in certain other countries

First Edition published in 2021

Impression: 1

Published in the United States of America by Oxford University Press
198 Madison Avenue, New York, NY 10016, United States of America

British Library Cataloguing in Publication Data

Data available

Library of Congress Control Number: 2021937887

ISBN 978–0–19–885026–7

Printed in Great Britain by
Bell & Bain Ltd., Glasgow

*To Dominic Burbidge, Fiona Gatty, Pete Jordan,
and Nikki Macmichael*

Preface

Why we wanted to write this book

When humans reflect on what matters most to them in life, they are probably thinking about an aspect of human flourishing. Most people want to live flourishing lives, and want those whom they love to flourish too. But many things make flourishing difficult in uncertain times. It is stressful to have expectations of being able to control one's destiny, only to find that one's best intentions are based on assumptions which prove to be unfounded. It can be like a person who has become used to controlling their room comfort though setting a precise thermostat finding that they must now adapt to fluctuating temperatures not of their choosing. Human flourishing has to be robust against uncertainties in our knowledge about the present and our predictions about the future. This book is based on the conviction that promoting human flourishing requires the best of scientific insight and the best of spiritual wisdom. For some readers these may seem to be improbable bedfellows. It might be thought that spiritual wisdom belongs to a bygone pre-scientific age and that there is now no place for it; conversely, it might be felt that science has little to offer in making life's most important decisions. We reject both positions. We see a great need for scientific insight to help tackle humanity's greatest challenges, and we also see a great need for spiritual wisdom to use the fruits of science well and to address questions which lie outside the self-limited scope of science. We believe that living well is facilitated by a harmonious respect for contributions from different realms of human inquiry and experience.

Before good choices can be made in life, it is necessary first to identify what a good outcome would look like. That means knowing what it means for humans to flourish. We reckon that there are three dimensions of human flourishing which cannot be separated but which can be distinguished for the purpose of considering them. The first is *material*, because humans cannot flourish without adequate water, food, shelter, and bodily health for themselves and their families and the wider community. The second is *relational*, because humans cannot thrive in isolation; we need to be with others and we have evolved to relate to others. The third is *transcendent*, because with only the material and the relational there is still something missing without which humans experience a kind of spiritual poverty. These dimensions are connected, because each of them can find expression through the other two.

Human flourishing is not suspended in mid-air, with no visible means of support, and subject only to the passing whims of intellectual and moral fashions. We identify three pillars which together can support a stable platform for human flourishing. These are *truth*, because humans cannot live well on the basis of lies; *purpose*, because humans need to know what they are here for; and *meaning*, because humans desire to lead meaningful lives. Humans do not flourish if they are living in ways that are false, aimless, or meaningless. Truth, purpose, and meaning each involve a judicious combination of objective reality and subjective response.

The uncertain times in which we live present changing contexts for human flourishing in which previous approaches need either to be redirected or at least applied afresh. We focus on three different instances of contemporary change in the world, intending our selection to be illustrative rather than exhaustive. There is a growing awareness of the unpredictability of life, which is revealed by the way that the social sciences are recognising more complex, and more morally load-bearing, models of human motivations and human interactions. There are changing patterns of religious commitment, with the rise of other faith traditions in historical Christendom and beyond, and talk of a post-Christian Europe sitting alongside talk of a post-European Christianity. And the pace of technological innovation is accelerating in fields as diverse as machine learning and gene synthesis, crying out for overarching principles to guide their use. We hope that case studies like these will bring to life how scientific insight and spiritual wisdom can together promote human flourishing in uncertain times.

A Beatles song of the 1960s, which from time to time enjoys a retro kind of revival, carries the refrain, 'All you need is love.' A word which covers too much can end up conveying rather little. By itself, asserting that love is sufficient does not get us far. Our hope is that by elucidating the dimensions of human flourishing and how it rests on robust pillars, we have shown how scientific insight and spiritual wisdom can work together for good. This needs to be, and we think can be, resilient against the vicissitudes of life. At its best, love does indeed provide the essential resource for human flourishing.

Who we want to thank for this book

The idea for the book arose from discussions in Oxford with Fiona
Gatty, Pete Jordan, Nikki Macmichael, Andrew Serazin, and Bonnie
Zahl, and subsequently also with Dominic Burbidge. Pete gave invaluable
comments on an early draft of each chapter. Our thinking about
human flourishing was stimulated and informed by interviews and
conversations with Anthony Aguirre, Anna Alexandrova, Terrence
Ascott, Israel Belfer, Sarah-Jayne Blakemore, Arthur Brooks, Sarah
Coakley, Paul Collier, Paul Davies, Simon DeDeo, Andrew Dilnot,
Michael Ebstyne, George Ellis, David Ford, Nidhal Guessoum, Peter
Harrison, Hermann Hauser, Rolf Heuer, Malcolm Jeeves, Tom McLeish,
Colin Mayer, George Monbiot, David Myers, Onora O'Neill, Martin
Rees, Beth Singler, Santiago Siri, John Stackhouse, Max Tegmark, Tyler
VanderWeele, Rafel Vicuña, Miroslav Volf, Justin Welby, and Adrian
Weller. Sonke Adlung and Giulia Lipparini are our unfailingly supportive
editors at Oxford University Press; Helen Reilly researched the pictures
and obtained the permissions; Charles Lauder Jr copy-edited the
typescript. The project was funded by a grant from Templeton World
Charity Foundation. The opinions expressed are ours, not those of the
foundation. We take the responsibility for them but not all the credit,
because countless people, starting with our parents and those who
came before, have helped us to become who we are and to think what
we think.

Contents

1

Dimensions and Pillars of Human Flourishing

Angela Ricker was born in 2004 in the living room of her parents' home, surrounded by her immediate family. No sooner had she arrived in the world than it was apparent that all was not well. Her uncle remembers receiving a distressed phone call from her grandfather, 'There's something wrong with the baby.' Babies with Down syndrome have an extra copy of their twenty first chromosome—a condition called trisomy 21. Angela had three copies of her thirteenth chromosome—this is called trisomy 13 or Patau syndrome. Angela's trisomy 13 affected almost every part of her body, from curled-in toes to unfused plates in her skull. It was this that immediately alerted the midwife to the need for urgent medical attention.

Angela never learned to see or hear normally, to walk or talk, or to feed or wash herself. Her family would never know what she understood or appreciated about her mother and father, her brothers and sisters, or her wider family. Although early on she responded to voice and touch, as she grew she seemed to recede further into an already distant and unknowable world. Was Angela flourishing?

Her uncle reckoned that you could not answer that question in isolation. If a test of a human community is how it cares for the most vulnerable like Angela, then the question is not whether Angela was flourishing *on her own* (she could not), but whether her presence led to the family and others flourishing *together*. That in turn suggests other questions, such as 'Who is helping Angela to flourish?' and 'Who is flourishing because of Angela?' Maybe even, 'How can those around her become the kind of people among whom Angela flourishes and who flourish with Angela in their midst?'

When Angela was ten years old, her grandparents celebrated their fiftieth wedding anniversary. Angela's siblings and cousins played music and games around her. Without belittling the huge emotional and financial costs of caring for Angela, her family found that her inescapable vulnerability in an astonishing way concentrated their attention and their love. Her uncle described it thus, 'In a centrifugal world where everything and everyone flees the demands of love, Angela was a centre of gravity, drawing us back to one another and to true life—the life that really is life, the life that money cannot buy, the life of making flourishing possible, at great cost and with great tears.'[1]

For Angela and those around her to flourish required the best of scientific insight—the medical science that enabled her condition to be diagnosed and underpinned all the equipment and treatment needed to keep her alive—and the best of spiritual wisdom—the wisdom that affirms that Angela is of value and inspires those around her to keep on loving her even when there is no evident response. Those resources needed to be applied in three connected dimensions: the material, her bodily needs; the relational, the caring by her parents and others; and what we shall call the transcendent, to describe what enabled her family to attribute dignity and value to Angela.

Angela died a year later. While she was still alive, her uncle posed two further questions about flourishing: 'What are we meant to be?', and 'Why are we so far from what we're meant to be?'[2] Answering such questions requires rich resources of scientific insight and of spiritual wisdom.

What do we mean by 'human flourishing'?

The notion of 'flourishing' may appear rather antiquated to some. In the Western tradition the term is most closely associated with Aristotle, who lived in the fourth century before the Common Era. But interest in human flourishing has been growing in recent years in academic, policy, and popular circles. In part, this is a reaction against measures such as 'average human lifespan' or 'per capita Gross Domestic Product (GDP)'. While such measures clearly tell us something about a nation's or a community's state of affairs, we all know that a long life is not necessarily a good life and the same is true of a life that is rich if measured only in monetary terms. What then is it that makes a human life a flourishing life? We start from the premise that each person is of equal dignity—a premise shared by many religious and non-religious traditions. As this book is about to go to press, the refrain 'Black Lives Matter' has revived an awareness of the inherent dignity and worth of every human.

Possibly since humans were first capable of asking the question, certainly since the dawn of history, humans have asked why we are here and what a good life entails. At different times, different answers have held sway. Nowadays, there are perhaps more answers proposed than ever. Much of humanity still finds the ultimate answers to meaning and purpose in religion. But in countries across the globe, secular views are widely held. In any event, whether religious or secular, individuals, communities, and governments still have to make decisions about what people want and need from life.

The notion of human flourishing is a useful concept within which to consider such questions—few would maintain that we want people not to flourish. The concept is sufficiently flexible that it can contain common-sense answers as well as ones that date back to the births of the world's major religions and the origins of philosophy, whether in the East or the West. In this book we therefore explore what is meant by human flourishing and see what it has to offer for those seeking after truth, meaning, and purpose. We hope that this book will enable readers to clarify what they want for their lives, for themselves, for their families and more widely. In our more optimistic moments, we hope that what we write will help some to lead more flourishing lives. We are not so naïve as to imagine that our writing will help those who, for example, are clinically depressed—and we are not attempting to write a self-help guide. But we do believe that at a time when most of us are bombarded with messages about what we should or should not do to live healthily, attain a work–life balance and find meaning, a careful consideration of the contributions of both scientific insights and spiritual wisdom to human flourishing can provide a new angle that many will find helpful.

We realize that not everyone is convinced that science (both the natural sciences and the social sciences) has much of value when considering questions of human flourishing, such as purpose and meaning. Equally, many hold that, important as such questions are, religions have nothing of value to contribute to them. We disagree with both these views. In different places within the book the extent to which we rely on scientific insights—including such social sciences as psychology and sociology within the scope of science—and spiritual wisdom varies. In some chapters one takes precedence, in other chapters the other. Across the book as a whole, both make major contributions and our hope is that someone sceptical of the utility of one or the other will, if not converted to our position, at least appreciate why we have included both and understand the contribution that each makes to our argument.

To anticipate the argument that we will develop, we maintain that the concept of human flourishing provides a valuable framework within which to consider the importance of satisfying people's yearnings for material goods, successful relationships and the hope that we can achieve and experience things that give us a sense of something greater than ourselves—the transcendent. We maintain that the transcendent is not discerned only within religions; for many, the arts, nature, wilderness, and a consideration of our place in the Universe are all instances of routes towards an appreciation of something beyond. At the same time, transcendence plays a particular role within religions and we will discuss aspects of transcendence that are opened up by a religious or spiritual outlook on life.

The material dimension of human flourishing

Words matter. The word 'material' can be understood in a number of ways. In philosophy the term 'materialism' refers to the view that nothing matters except for matter and such physical concepts as energy and waves. One of us is based in the Oxford Department of Materials. A scientist who works in physics is a physicist, but to describe a scientist who works in materials as a materialist might be seriously misleading! By the 'material dimension to human flourishing', we mean those aspects of human flourishing that are to do with such things as having enough to eat, access to clean water, enough sleep, reasonably good health, somewhere that one considers to be one's home and in which one feels safe, and enough money not to be endlessly worried by financial matters.

It might immediately be objected that we are beginning rather to stretch the everyday understanding of 'material'—for example, in our inclusion of 'somewhere that one considers to be one's home and in which one feels safe'. Our reason for having a broad conception of the material dimension to human flourishing is that we don't want simply to erect a 'straw man' definition of the term which allows it almost effortlessly to be knocked down, leading to the conclusion that the material dimension on its own is insufficient. We do think that even our rather broad understanding of the material dimension to human flourishing provides an inadequate conceptualization of human flourishing. One of the rather sad, in our view, features of much of modern life, including too many education systems, is that one can arrive at adulthood thinking that the material is all there is to life.

However, this is to get ahead of ourselves. Before we critique the notion that the material dimension is enough for human flourishing, we need to acknowledge that for many people the way we have characterized it—such things as having enough to eat, access to clean water, enough sleep, reasonably good health, somewhere that one considers to be one's home and in which one feels safe, and enough money not to be endlessly worried by financial matters—sounds like a utopian dream. Even in peace time there are many hundreds of millions of people across the globe who do not enjoy such basic comforts. And when we add in the effects of wars and other conflicts, we are talking of many more.

So the material dimension does matter. Indeed, if we just think of having enough sleep and enjoying reasonably good health, things may be getting worse in many countries. Modern life for many of us, as we live in an increasingly 24/7 wired world, means that it's all too easy to deprive oneself of sleep, striving to keep connected to our social networks for just another half hour. And then, while almost all countries have seen startling improvements over the past century in life expectancy, this does not seamlessly translate into greater human flourishing for all. People may be living some 30 years longer on average than they did a century ago but around ten of those additional years are often ones of poor health. The average person nowadays spends longer, especially towards the end of life, in poor health than their ancestors did.

The relational dimension of human flourishing

Most of us enjoy the company of others, even if there are times when we may prefer to be on our own. We value family and friendship, even though we all know that family relations can be painful and we may fall out with our friends. But even these apparent objections to the value of relationships show how significant they are—they can go wrong and damage us as well as go well and help us to flourish.

For all of us, our initial closest relationship is with the woman in whose womb we begin our post-conception life. Throughout our lives we carry genetic material from two individuals (setting aside biotechnological interventions such as treating mitochondrial disease using gene therapy) but for some nine months or so we rely on the biological environment that one of them provides. This first relationship is an important one. If all goes well the baby emerges at birth having developed from a single fertilized cell into a newborn, typically weighing several kilograms, able to breathe on its own and begin its post-partum development through to adulthood.

Sometimes, though, matters don't go as well before birth as they should, and not only for genetic reasons. For example, if a pregnant woman's diet is low in folate (vitamin B_9) in the first few months of pregnancy, the newborn may have neural tube defects, which can have adverse lifetime consequences. Moving from shortage to excess, if a pregnant woman has too high an intake of alcohol or smokes cigarettes (or pretty much anything else), then there can be damage to her developing child. Fetal alcohol syndrome can result in permanent brain damage, with consequent harms to educational attainment and general intelligence as well as other problems including motor coordination. While mental impairment should not be equated with a life that is less worth living, no parent wants a child to have fetal alcohol syndrome.

The relationship a mother has with her unborn child illustrates how the distinction between the material and relational dimensions of human flourishing may not always be clear. In one sense, all the adverse consequences in the preceding paragraph could be said to provide evidence for the material dimension—too little folate, too much alcohol or cigarette smoke. It is because the unborn child sits within his or her mother that we can also consider the effects as illustrating the relational dimension.

After the baby is born, the primary benefit of feeding, whether by breast or bottle, comes from the nutritional, which is clearly material. But feeding for many of us—especially for a mother (or father, though perhaps typically to a lesser extent) with her newborn child—is also about relationships. This is a lesson from the great 1987 Danish film *Babettes gæstebud* (*Babette's Feast*), based on a story by Isak Dinesen (aka Karen Blixen). Spoiler alert: Babette is a refugee who uses lottery winnings and her extraordinary culinary expertise to create a meal that heals a damaged community.

For a mother and baby, feeding plays an important part in their relationship. New mothers are bombarded with feeding advice from all corners. The consensus nowadays seems to be that whilst breastfeeding can have various health benefits over bottle feeding (it is still not really possible to replicate in formula milk all the constituents of breastmilk, which also plays a role in the development of the immune system and lowers the likelihood of Sudden Infant Death Syndrome (cot death)), a mother can establish a good relationship with her baby whether she breastfeeds or not.[3] We would like to think that this observation can be extrapolated to bottle-feeding by fathers.[4] It is because of the importance of this bonding that some nursing parents choose to put away their phones while feeding.

The psychoanalyst Wilfred Bion introduced the term 'reverie' (from the French for 'dream') to describe what can happen between mother and baby when feeding is going well—which, of course, it doesn't always, for a range of reasons, sometimes to do with the baby, sometimes to do with the mother and often to do with their relationship, even if only their relationship at a particular point in time. When feeding goes well, the baby, with its immature mental structures, somehow senses that its mother can contain any anxieties it has. The baby can therefore relax and concentrate on feeding and on its relationship with it mother. At the same time, the mother too may enter a state of reverie. This experience, for all that it is a natural and not uncommon experience, may suggest to the mother an aspect of the transcendent. There can be a depth to the experience that is beyond the everyday, much as some poets talk of the capacity of nature to take us out of ourselves. We have more to say about transcendence in the next section.

Beyond the mother–baby relationship, most of us would affirm the importance of good relationships between people. This applies to dyadic relationships—as in a marriage or between a parent and a particular child or between two friends of any age—and also to relationships within a community, whether one is thinking of the relationships within a family, a team, a congregation, a neighbourhood, or any other group that is not so large as to be anonymous.

Just how important relationships between people are for human flourishing we will examine in Chapter 3. For some people, relationships with non-human animals can play a major role in helping to maintain their quality of life. Distinctions can be made between pets, companion animals, and service animals. Pets are usually domesticated animals (cats, dogs, horses, certain bird species, etc.) that by virtue of their domestication are easy to keep and generally get on well with most

Figure 1.1 Companion animals can make an important contribution to people's flourishing.

people. Individuals with some sort of disability may have a companion animal, which has no particular training for its role, or a service animal, which does. A companion animal, like a pet, can provide company, enjoyment, and psychological support (Figure 1.1). A service animal, such as a guide dog for someone who is blind or visually impaired, does more than this—and is usually trained to pay as little attention as possible to members of the general public, so that it can focus on its job.

The transcendent dimension of human flourishing

For most of us, there are times when aspects of life seem to go beyond, to transcend, the quotidian or the mundane. Nature, music, poetry, and the other arts can transport us beyond ourselves. Many creative people, whatever their discipline, may feel as if at least part of what they are creating comes from outside themselves, is given to them. An awareness of the transcendent can happen when we are alone or in group situations, whether in singing, in dancing, in certain sporting activities, or in worship.

But what do such highfalutin statements mean? Feelings like these, which in groups can be at least partly triggered by endorphin release, may serve some evolutionary function, perhaps in terms of binding people together; there is a growing scientific literature on this. Is that the whole story? Or does the way we react to great works of music, stunning scenes in nature or the birth of our own children require a deeper explanation?

We do not want to advocate some sort of cheap argument here from such experiences to belief in a transcendent being. Neither of us thinks that that sort of argument works. At the same time, while we fully recognize that there can be rich secular interpretations of such phenomena, we are also entirely comfortable with the suggestion that some experiences can be more than this, that they can link us to an awareness of the divine.

At this point it seems sensible to say a bit more about each of us. We were both educated in the natural sciences and one of us, Andrew Briggs, has remained in them to this day while the other of us, Michael Reiss, soon migrated to the social sciences. Each of us has a Christian faith that is important to us and shapes how we try to live our lives. For us, therefore, the transcendent dimension of human flourishing is as significant as the material and the relational. For most of our professional lives each of us has sought to elucidate how our occupation fits within our religious convictions.

Since Christianity is the tradition which we know best, when we consider religion we focus on Christianity. At the same time, we have tried not to be too parochial, indicating when our observations apply to religions in general and, in places, drawing on religious traditions other than Christianity. We hope that readers will be able to apply the principles which we set out to wherever they are coming from.

The Methodist theologian Frances Young has spent most of her adult life struggling with the practical and theological issues arising from the birth, in 1967, of her son Arthur. Arthur has severe physical and mental disabilities, is unable to speak and has always required a great deal of care. In her book about Arthur,[5] she concludes with a chapter titled 'Arthur's vocation', which provides insights about both the relational and transcendent dimensions of human flourishing.

Arthur, and others like Arthur, including Angela Ricker with whom we started this chapter, enable a shift from individualism and competitiveness to community and mutuality. As Young puts it:

What really makes us human is the capacity to ask for help, and that
challenges modern claims to autonomy, as well as our individualism and
success-values. The spirituality of the L'Arche communities has much to
teach us about the presence of God in the everyday experience of living
with persons who have learning disabilities. It's important to highlight
the mutuality of this relationship. It's not a matter of doing good, or
patronizing charity, but of receiving as well as giving, according dignity
to the other person by receiving from them. The fruits of the Spirit,
according to St Paul, are love, joy, peace, patience, kindness, generosity,
faithfulness, gentleness and self-control (Gal. 5.22). It is in community
with persons who are limited in their competence and capacity, at least
compared with most of us, that we often best discover these deeper
values.[6]

Young concludes her book with the thought that her son functions as a
religious minister, who reveals to us something about who we are and
reminds us that in worship believers enter the wordless praise along
with all of creation.[7]

Different people are likely to react to the suggestion that there is a
transcendental dimension to human flourishing in different ways. We
hope that many readers find the notion intriguing, possibly attractive,
even if it is not one to which they may have previously given a great
deal of thought. The transcendent dimension of human flourishing,
like the relational and material dimensions, is not without support. We
identify three robust pillars: truth, purpose, and meaning.

The pillar of truth

We take it as axiomatic that a flourishing life will be built on truth.
Pilate asked 'What is truth?' and, whatever he meant by this,[8] the
question remains an important one. In the context of a post-positivist
hangover,[9] it is easy to assume that the only truth that matters is
empirical truth of the sort that can be used to establish whether a
statement such as 'Gold is a metal that does not tarnish' is true or not.
An assertion like this about the physical world falls within the domain
of the natural sciences. How would a scientist go about establishing
whether it is true or false? First, it would be necessary to be precise about
the various terms. 'Gold' causes no problems—the word clearly refers
to the chemical element with an atomic number of 79—and 'metal' is
reasonably clear-cut (though non-scientists may be surprised to be told

that mercury is a metal as there is no requirement for a metal to be solid at room temperature) but 'tarnish' is a bit more problematic. It is an everyday word and everyday words often lack the precision that scientists attach to words—for example, to a physicist, the words 'energy', 'work', and 'force' each have precise and distinct meanings, which they lack in day-to-day conversation. In the case of 'tarnish', the word principally refers to the product of a chemical reaction between a metal and either oxygen or sulphur dioxide. Then, having clarified precisely what is meant by 'Gold is a metal that does not tarnish' it would be necessary, either through experimentation or some other objective method, to establish whether gold does indeed tarnish, or whether this is only the case for a substance that contains gold in a mixture (as gold in jewellery almost always is—even 22 carat gold has 8.3 per cent non-gold metals like silver, zinc, and nickel), rather than when it is pure.

But there are other ways of establishing truth, in addition to those used in the natural sciences. Mathematicians establish truth by ensuring that assertions that fall within the domain of mathematics are consistent. Many of us may remember from our school days the threefold classification of mathematics into arithmetic, geometry, and algebra, but there is more to mathematics than this. Mathematicians are fascinated by patterns and while there is no universally agreed definition of the subject, mathematics is widely agreed to include such things as the theory of knots and game theory. Mathematicians arrive at their conclusions through the use of proofs—all it takes for a purported proof to be invalidated is for it to be shown to have one inconsistency or a missing step that cannot be filled in. Something of what it is like to be a world-class mathematician is captured in Simon Singh's account of how Andrew Wiles proved Fermat's Last Theorem, a mathematical problem that had baffled mathematicians for over 350 years.[10]

Truth can be found in other domains of knowledge—in history, in aesthetics and in moral philosophy, for instance. There is a joke that goes 'If Henry VIII had six wives, how many wives did Henry IV have?' The humour relies on the appreciation that anyone (perhaps a young child) who seriously answers 'three' has failed to understand how both mathematics and history work.

But there are deeper questions about truth that the sciences, mathematics, history, aesthetics, and moral philosophy cannot answer.

In *The Republic*, Plato presents his allegory of the cave. Plato has Socrates describe the lives of people (perhaps us!) who live their lives in a cave where their perceived reality consists only of shadows projected onto a wall from a fire behind them. From within a system it can be difficult to imagine what the system looks like from the outside. It's a bit like the story of two embryos debating whether there is life after birth. One asserts that there is; the other maintains that there isn't and that stories of life after birth—of embryos entering though a tunnel into the light of a new world—are wish fulfilments. After all, what embryo has every come back from birth to convince other embryos of this second life?

The pillar of purpose

When Charles Darwin was considering whether to propose to his cousin Emma, he listed the advantages *of not marrying*, including:

> Freedom to go where one liked—choice of Society & *little of it*.—Conversation of clever men at clubs—Not forced to visit relatives, & to bend in every trifle.—to have the expense & anxiety of children—perhaps quarelling—**Loss of time**.—cannot read in the Evenings—fatness & idleness—Anxiety & responsibility—less money for books &c[11]

and also the advantages *of marrying*:

> Children—(if it Please God)—Constant companion, (& friend in old age) who will feel interested in one,—object to be beloved & played with.——better than a dog anyhow.—Home, & someone to take care of house—Charms of music & female chit-chat.—These things good for one's health.—*but terrible loss of time.*—
>
> My God, it is intolerable to think of spending ones whole life, like a neuter bee, working, working, & nothing after all.—No, no won't do.—Imagine living all one's day solitarily in smoky dirty London House.—Only picture to yourself a nice soft wife on a sofa with good fire, & books & music perhaps—Compare this vision with the dingy reality of Grt. Marlbro' St.
>
> Marry—Mary—Marry Q.E.D.[12]

More generally, we can say that the purposes of a marriage include companionship for the couple, a stable basis within which to bring up children and a socially sanctioned mechanism by which two people can begin a new life together. As an Anglican statement of purpose of marriage expresses it: 'Marriage is given, that husband and wife may comfort and help each other, living faithfully together in need and in plenty, in sorrow and in joy. It is given, that with delight and tenderness they may know each other in love, and, through the joy of their bodily union, may strengthen the union of their hearts and lives. It is given as the foundation of family life in which children may be born and nurtured in accordance with God's will, to his praise and glory.'[13]

To a reductionist evolutionist, the purpose of life is to produce more life, life that is as closely related as possible. The past forty years have seen an enormous growth in the disciplines of behavioural ecology and evolutionary biology, with a particular focus on cases where organisms appear to engage in behaviours that contradict this simple dictum. A classic case is altruism—cases where organisms help one another in ways that go beyond what might be regarded as unproblematic instances of parents assisting offspring. Darwin himself wondered about the evolution of sterility in the social insects. In many species of ants, bees, termites, and wasps, many individuals—indeed, typically the large majority of them—never attempt to reproduce, instead serving the colony as a whole. Darwin realized that what such individuals are, in a sense, doing is to reproduce vicariously via others in their colony.

Nowadays we realize that the story is a bit more complicated—there are evolutionary battles within a colony as the various individuals do not all share identical interests—but the fundamental insight of Darwin holds good. This type of activity is nowadays named 'kin selection' as individuals are, effectively, reproducing via their kin (e.g. their siblings) rather than directly. Another mechanism by which altruism can evolve is through reciprocal altruism when one individual helps another individual (who may not even be in the same species) with the expectation (though this is not to imply any conscious awareness) that the time will come when such behaviour will be reciprocated and the altruist thus paid back.

When most of us wonder at a deeper level what we should do with our lives, consideration of how natural selection of random mutations of genes might best advance our interests is not generally uppermost in our minds. Part of being human is that we are able to go beyond the forces of evolution in a way that may be unique to our species. It is hard for people to flourish if they feel that their lives, indeed the Universe more generally, lack purpose. A key issue here is what is involved in finding a purpose. At one pole is the view that the Universe has no ultimate purpose for us; each of us needs, if we give any thought to the matter at all, to invent a purpose for our life. At the other pole is the belief that the Universe has a defined purpose for us, which each of us needs to find. This latter perspective is found in most of the world's religions, though they differ greatly as to the origin and nature of this purpose. In any event, whether invented or found, whether in family, in friends, in art, in politics, in patriotism, in religion, or elsewhere, the discernment of purpose can contribute greatly to human flourishing.

The pillar of meaning

To a certain extent we can find meaning in the goods we acquire and the other components of the material dimension of human flourishing but we are more likely to find meaning in the skills we develop and in the things we create. When we learn to read, to ride a bicycle, to swim, or to get around using a foreign language, the learning itself as well as what is learnt can be meaningful. Indeed, one of the healthy things about learning any new skill as an adult is that this can remind us how poor we initially are at something with which we are unfamiliar, how hard learning can be and, if all goes well, how satisfying it can be to see ourselves progressing—the first time, for example, we don't have to look at the keyboard when typing or when we suddenly realize that we are able to understand some of what is on a foreign menu.

Creating something can be a deep source of meaning. This is not only the case for the few of us who paint well or write poetry that someone else wants to read. It also applies when we cook meals or tend gardens—so long as such activities are not drudgery, which they can become if we are tired or in pain or if no one appreciates our efforts.

The clause 'if no one appreciates our efforts' indicates the importance of relationships to our location of meaning. Appreciation from a human being for almost any activity, however large or small, can imbue it with additional worth. We can even value appreciation from non-humans—a dog wagging its tail at the prospect of a walk or a cat purring when stroked—and the relational dimension to meaning, while aided by appreciation (which makes the relationship a two-way one), does not require it (think of a parent going to check that their child is asleep).

Meaning is especially likely to be found in our deepest relationships. These can be friendships though marriage (or its cultural equivalents) is for many people more likely to be key, along with other deep family relationships, including those that parents have with their children.

We may have to dig deeper for meaning in circumstances where our material and relational needs are not satisfied. Those with a deep religious faith are likely to find meaning in their faith, always remembering that in religions where immanence is taken seriously, part of our relationship with God manifests itself in our relationships with those around us and in the significance of the physical environment. In awful situations—whether ones that almost all of us face at some time, such as the death of a loved one, or ones that, thankfully, few of us experience, such as starvation or torture—some people are able still to locate meaning, even if only in how they exercise the few choices that they are still able to make.

Limits to predictability

It is sometimes supposed that the natural sciences, and in their wake the social sciences, are all about making precise predictions—that the more information we have, the more exactly we know what will happen. This view came into prominence after the towering achievements of Sir Isaac Newton, who realized that the same laws of motion that describe how objects fall to the ground also describe how the planets in our Solar System revolve around our Sun, and the Moon around the Earth. A clockwork-like model of the Universe came to predominate in which it seemed that a precise knowledge of the present state of affairs would enable one to predict events far into the future.

There are areas where this is indeed the case. When we are dealing with situations involving massive objects, like the movement of the Sun and the planets, where gravity is the determining force, we can often predict events far into the future (Figure 1.2). We can confidently predict that there will be (or, by the time you read this, will have been) a total solar eclipse on 4 December 2021 and another on 8 April 2024. Astronomers can predict the date of such events hundreds of years ahead and, to a precision of a second, when they will start and end.

But even in physics there is much that is uncertain. Take something as well known as radioactive decay. We cannot predict when a radioactive atom will decay; we can predict the probability of it decaying

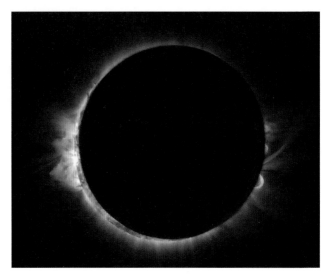

Figure 1.2 We can predict the timings of solar eclipses far into the future with great precision. But much in the natural world is far less predictable.

over a certain time period—but that is as far as it goes. Furthermore, the advent of chaos theory has enabled us to realize that, contrary to early aspirations, there are many features of the natural world that we will never be able to predict with useful accuracy—the weather on a given day a year from now being an example.

None of this, of course, means that the world doesn't obey the laws of physics—just that the laws of physics are somewhat more complicated and much more surprising than was once thought to be the case. Nor does it mean that the world is irrational—rationality and irrationality are features of human minds.

At one point, many social scientists thought that there were laws that, once we discovered them, would enable us to predict the behaviour of humans with perhaps the same reliability as we can predict the behaviour of inanimate objects, thanks to the laws of physics and chemistry. This presumption has faded somewhat, though it keeps on reappearing, often as a new science or technology develops. So genetics is not infrequently equated with determinism and the advent of neuroimaging has led some to presume that we will soon be able to predict how humans behave.

There is a growing push back to such presumptions. The term *Homo economicus* was introduced in the nineteenth century to characterize humans as beings who are consistently rational and act in their own interests. At one time loved by economists—these sorts of assumptions make mathematical modelling much more feasible—the problem with such a characterization is that it fails on empirical grounds.[14] Almost no one is consistently rational and, thankfully, almost no one invariably acts in their own interests. We may be rational and self-interested much of the time but our actions are often better predicted by our values, which, while they need not be irrational, often vastly exceed what is simply rational and selfish. Many of us at times show compassion and generosity, even with no realistic chance of these benefitting us now or in the future.

One of the important implications of the world in general and humans in particular being less predictable than has often been supposed is that the future is more open-ended. The lives we lead and the choices we make make a difference. We can promote flourishing, both for ourselves and for others, or we can make it less likely.

Patterns of religious commitment

These major shifts in how we understand the predictability of the natural world and the rationality of human action have come about at a time of tremendous changes in religious observance. Whilst worldwide some 90 per cent of people profess a faith in God, religious observance, as measured by stated beliefs or documented practices, is falling in many countries, albeit remaining stable in most and rising in a few. In some countries that score highly on such measures of human flourishing as per capita GDP or average human happiness, religious observance is low and falling.

We do not find it particularly surprising, given our multidimensional approach to human flourishing, that countries with high per capita GDPs or measures of human happiness, such as the Nordic countries, have low levels of religious observance. We might expect that if material or relational measures of flourishing were high, religion might be less important—it is a widespread notion, and one for which there is some supporting evidence, that people are more likely to manifest religious practices such as prayer or attendance at worship when times are tough.

The decline of religious observance in much of the West has given rise to what Grace Davie has characterized as 'believing without belonging', i.e. those who have some degree of belief without being a regular part of any worshipping community.[15] To this we can add the notion of 'belonging without believing', of whom no doubt there have often been many but whose numbers may be decreasing where there are fewer social pressures favouring regular attendance at worship.

The place that religion plays in human flourishing is profound and complex. For the atheist, religion fails the 'but is it true?' test which we examine in Chapter 5. The many reported and positive correlations of religiosity with measures of self-reported happiness are then seen as evidence of self-deception. For those with a religious faith the correlations are not unexpected. Furthermore, the religious believer is not surprised that religiosity also correlates positively with more objective measures such as physical health and longevity, though the relationship between religiosity and mental health is less clear.

There are deeper questions about religion and flourishing than whether people with a religious faith live longer or report that they are happier. In Chapter 4 we examine whether a spiritual framework for one's life provides access to a transcendental dimension that a secular appreciation of beauty in, for example, nature, art, and music cannot. In Chapter 9 we examine the implications of changing patterns of religious observance.

Human flourishing in an age of technology

Human technology predates the origins of *Homo sapiens*. Our ancestors were making stone tools millions of years before our species had evolved. The relationship between humans and technology is bidirectional. It is not only that our ancestors changed stones through chipping away at them so that they could be used for arrowheads, for chopping or as querns. Such stones, and tools in general, change us. They favour individuals who can make them and use them well. They contributed to important changes in our anatomy and behaviour and enabled us to increase in numbers and to spread to new geographical areas.

Despite concerns that robots might take away people's jobs, their expense and initial inflexibility meant that the movement to robots from humans, in things like car assembly lines, was quite slow, giving time for humans in the industry to move to other jobs. However, robots are becoming more affordable and more versatile with the advent of machine learning. Calculations suggest that a number of jobs in which large numbers of people are currently employed will soon no longer exist. For example, if autonomous lorries really do become a reality, many of the 3.5 million professional truck drivers in the USA will need new jobs. Not all of them will want to join the service industries. Besides, the use of robots in the service industries is likely also to increase substantially. Whether we really will have robots taking our orders in restaurants is unclear—though the prevalence of systems where one simply types one's choice into a tablet rather than engaging with a human being is increasing. However, robots are likely to become more important in social care (Figure 1.3). There is therefore a risk that elderly people in care homes might enjoy less human contact. On the other hand, a robot may be quicker to help one go to the toilet in the middle of the night—and some people may prefer to be helped in this by a robot than by another human being.

Machine learning is one of the fastest growing fields in computer science. The ability of machines to outperform humans is being demonstrated in a growing range of applications, from board games to medical diagnosis. We can expect to see increasing benefits. And who, even before the 2020 pandemic, would want to be without the power

Figure 1.3 Some people, but not necessarily everyone, will welcome the increasing role that robots are likely to play in social care in certain countries.

of social connectivity and online shopping? But there lies the danger. Some of the best minds in the world are being paid some of the highest salaries in the world to target influence at each user, starting with all the devices at their disposal to maximize the time that vulnerable young people spend looking at their phone screens, and thence having their preferences manipulated both as consumers and as citizens. Flourishing is more than ever being controlled by large commercial interests, which are not necessarily aligned with an individual's conscious values.

Enhancement in humans exists in many forms. We can start with the uncontroversial—the education of a child by its family and in school. Education is all about enhancement; we talk about a learner 'fulfilling their potential' though it is often unclear whether there really are such limits and, if there are, what they are. Preventative approaches in medicine, such as inoculations, are all about enhancing the defence mechanisms of individuals so that they are resistant to polio, chickenpox, whooping cough, measles, mumps, rubella, and now coronavirus. Somatic gene therapy is emerging from the laboratory and into the clinic. Here, alterations are made to the DNA in the cells in the body that do not give rise to eggs or sperm. The intention is to correct those diseases or conditions that result from mutations in our DNA, such as sickle-cell anaemia and β-thalassaemia.[16] In all these cases, if safety concerns are met and people consent to procedures, the contribution to human flourishing is pretty clear-cut. More problematic are cases—hypothetical at present but perhaps not for long—where we are not talking about restoring gene function to normality or enhancing the body's capacity to deal with infectious diseases but things like using somatic gene therapy or pharmacological agents to enhance a person's ability to learn more quickly. Would this be a good thing, or would it give such individuals an unfair advantage? Who will decide whether and how to use such capabilities should they become available? Will they diminish or enhance human flourishing?

We live in uncertain times. When we started to write this book, we little thought that we would finish it in a world disrupted beyond any comparable experience by the COVID-19 pandemic. When you read it, other changes may be uppermost in your mind. By taking these case studies of changes in uncertainty and predictability, in patterns of religious commitment, and in technologies, we seek to illustrate how the foundations of human flourishing can be applied to the dimensions of human flourishing. We hope that these worked examples will bring to life what might otherwise seem ethereally abstract. Human flourishing is all about how humans actually flourish, now and in the future, however uncertain the future may be.

Organization of the book

The rest of the book is divided into three parts, each consisting of three chapters, plus the conclusion. Before each part we have an introduction to that part. Our first part is about the three dimensions of flourishing that we have identified: material, relational, and transcendent.

Notes

1 Crouch, A. (2016) *Strong and Weak: Embracing a Life of Love, Risk, and True Flourishing*. Downers Grove, IL: InterVarsity Press, p. 185.

2 Ibid., p. 9.

3 Hairston, I. S., Handelzalts, J. E., Lehman-Inbar, T. and Kovo, M. (2019) Mother–infant bonding is not associated with feeding type: A community study sample, *BMC Pregnancy Childbirth* **19**, 125.

4 Machin, A. (2018) *The Life of Dad: The Making of the Modern Father*. New York: Simon & Schuster.

5 Young, F. (2014) *Arthur's Call: A Journey of Faith in the Face of Sever Learning Disability*. London: SPCK.

6 Ibid., p. 143.

7 Psalm 148.

8 Wright, N. T. (2019) *History and Eschatology: Jesus and the Promise of Natural Theology*. London: SPCK, p. 239.

9 O'Neill, O. (2013) Science, reasons and normativity, *European Review* **21**, S94–9.

10 Singh, S. (1997) *Fermat's Last Theorem: The story of a riddle that confounded the world's greatest minds for 358 years*. London: Fourth Estate.

11 Darwin on marriage, Darwin Correspondence Project, Cambridge University Library, DAR 210.8:2. Available at https://www.darwinproject. ac.uk/tags/about-darwin/family-life/darwin-marriage. Darwin misspelled 'quarrelling'.

12 Ibid Darwin misspelled the first repetition of 'Marry'.

13 Church of England (n.d.) *Marriage Service*. Available at https://www. churchofengland.org/prayer-and-worship/worship-texts-and-resources/ common-worship/marriage#mm100.

14 Sen, A. K. (1977) Rational fools: A critique of the behavioral foundations of economic theory, *Philosophy & Public Affairs* **6**, 317–44.

15 Davie, G. (1990) Believing without belonging: Is this the future of religion in Britain? *Social Compass* **37**(4), 455–69.

16 Ledford, H. (2020) Quest to Use CRISPR against disease gains ground, *Nature* **577**, 156.

PART I

DIMENSIONS OF HUMAN FLOURISHING

Overview

Can there be a topic more vast than human flourishing? Almost everyone we know cares about human flourishing, whether or not they use that term. Family roles, not only as spouse or parent, should be about promoting flourishing. Many professions, whether as teacher, pastor, solicitor, or doctor, are explicitly about promoting flourishing, and at their best so is every other way of earning a living and contributing to society through voluntary or employed service. Humans have been thinking about what it means to be human since probably before writing was invented, and certainly ever since, from Aristotle writing about εὐδαιμονία, traditionally translated as 'happiness' but now more accurately paraphrased as 'human flourishing', to the Hebrew Psalmist asking God, 'What is man that you are mindful of him?'

In this book we identify three dimensions of flourishing, which we designate the material, the relational, and the transcendent. The material dimension has to be the starting point, because it is hard to enjoy the other dimensions of flourishing if you lack the basic requirements of water, food, shelter, and health care. And yet material prosperity by itself is unsatisfying, whether at the basic level required for survival or in the superabundance of the rich. Solitary confinement is a severe punishment. As the Hebrew story of the Garden of Eden recognized, 'It is not good for the man to be alone.' Humans are social animals; we need others. And yet, when all our material and relational needs are met, there still seems to be an appetite for something more. This can manifest itself in different ways at different times for different people; it may be enjoyment of a sunset, or a painting, or a poem, or a symphony, or a cathedral, or even a scientific discovery, in which we recognize a further reality which we can engage with even if we cannot pin it down. We call this the transcendent dimension.

These three dimensions may be distinct, but they are not isolated from each other. The closer we look, the more we find that they are entangled. That is why there are other possible descriptions. For example, flourishing can be broken down into economic, physical, and psychological domains. The psychological domain includes happiness, which is affective, involving a preponderance of positive emotion, and life satisfaction, which is cognitive with consideration of more objective achievements. Six domains of flourishing which have been proposed as likely to be widely accepted by most people in most cultures are:[1]

1. Happiness and life satisfaction
2. Mental and physical health
3. Meaning and purpose
4. Character and virtue
5. Close social relationships
6. Financial and material stability

To some extent 2 and 6 map onto our material dimension, 5 is a refinement of our relational dimension, and 3 corresponds to conscious engagement with what we shall address in Part II of this book. The list deliberately excludes domains of flourishing with which not everyone would agree, such as those pertaining to specific philosophical or religious traditions, which might belong to our transcendent dimension. Nevertheless, one can imagine that the transcendent dimension has the potential to impinge on each of these six domains.

Another kind of analysis identifies three criteria of life: life going well, life feeling well and life doing well. Life going well has environmental components: do I have enough food, shelter, clothing, health care, education? Life feeling well has emotional components; it requires physical and mental health, but it is more than health alone and usually involves good relationships. Life doing well is about me as a responsible person: what goals am I setting myself implicitly or explicitly, and what am I doing to deliver them? It can be that for a time all three aspects are in harmony. But there can also be times when there are conflicts between them. In our better moments most of us probably hope that life doing well would then trump the other two.[2]

Whatever breakdown one employs, the different elements can be distinguished but they cannot be separated. For example, if I extend life going well to my family, then part of doing well may involve providing for life going well for my spouse and my children, and possibly my parents and other relatives. If I extend life feeling well to my community, then part of doing well may involve serving the community in practical voluntary capacities. Although measures of human happiness and well-being often start with individual self-reporting, they can be extended to the flourishing of the communities in which those individuals are embedded.

We shall see in the following chapters how strongly the material, relational, and transcendent dimensions of flourishing are connected. Although it is theoretically possible for an individual to be self-sufficient in their food production, few in this planet actually are, and much of the evolutionary development of *Homo sapiens* has been towards ever greater cooperation.[3] Just imagine trying to make your own ball-point pen, let alone your own smart phone. We need each other for our material well-being. But we also use material means to build relationships. That is why we give presents. Parents build relationships with children, in part by providing them with food, shelter, and clothing, and then books and bicycles and whatever. So the interaction between the material and the relational dimensions of flourishing is two-way.

The entanglement between the transcendent and the other two is even stronger. A transcendent sense of awe and wonder may be inspired by the physical appearance of a starry sky, however elusive that may be in light-polluted urban environments. Transcendent experiences of poetry, art, and music are mediated by the physical experience of the word, the canvas, and the melody. They are often also relational; enjoyment of a creation is enriched by enjoyment of the creator, whether through personal knowledge of the evening's performer or through the adoration of the doting parent for their child's latest drawing. A moral imperative common to almost all world religions is to do for others what we would like them to do for us.[4] George Bernard Shaw cynically rejected this on the grounds that their tastes may not be the same,[5] but it is not difficult to make allowance for personal preferences. Loving your neighbour as yourself is relational, and it is almost always material too, because time and again we express our love through material means, whether it be cooking a meal or mending a bicycle. Within the Christian tradition the material, the relational, and the transcendent are fused through the incarnation, whereby, as John expressed it, the founding principle of the Universe became physically embodied in space and time.[6]

The three chapters of Part I are devoted to the dimensions of the material, the relational, and the transcendent. Our attempt to disentangle these three will unavoidably be only partially successful. On the way, we shall seek to enrich our exposition of what it means for humans to flourish.

Notes

1 VanderWeele, T. J. (2017) On the promotion of human flourishing, *Proceedings of the National Academy of Sciences*, **114**, 8148–56.
2 Volf, M. and Croasmun, M. (2019) *For the Life of the World: Theology that makes a difference*, Grand Rapids, MI: Brazos Press.
3 Henrich, J. (2017) *The Secret of Our Success*, Princeton, NJ: Princeton University Press. Christakis, Nicholas (2019) *Blueprint*, Boston: Little, Brown Spark.
4 Matthew 7:12.
5 Shaw, G. B. (1903) *Maxims for Revolutionists*.
6 καὶ ὁ λόγος σὰρξ ἐγένετο, John 1:14.

2

The Material Dimension

Poverty

In 2005 the UK newspaper *The Guardian* featured ten newborns in various countries across Africa. Five and then ten years later, the paper returned to talk to the children and their parents.[1] One of the children was David Lewis Dieumerci in the Democratic Republic of the Congo. By 2015 he had been going to a Catholic school a short walk from where he lives with his parents and three siblings. Maths classes are his favourite. The bad news is that he rarely gets to go to them. His father, Jean Mtoko, has been out of work for as long as anyone can remember. His mother, Ngosia Nzinga, had recently had surgery to try to recover vision in her right eye. The operation cost US$600 for which the family scraped together US$150 as a down payment. That left little money for school fees, which cost US$50 a term. And the rules are strict: no payment, no class.

There has been progress in the Congo. The rate of economic growth in the country is among the fastest in the world, driven by a booming mining sector. There are new roads, airport terminals, and trains. Some key development indicators have improved: the under-five mortality rate has dropped from 156 deaths among every 1,000 live births in 2005 to 119 in 2013. Despite a few brushes with malaria, David has been in good health, his parents say.

But in Makala, where David and his family live, there are no good jobs to be had. The lone light bulb in David's house flickers on, unpredictably, in spurts of an hour or two at a time. The government dreams of building the largest hydroelectric dam in the world near the end of the Congo River's long journey to the Atlantic, yet only about 10 per cent of the country's roughly 70 million people currently have access to electricity. The rubbish hasn't been collected in years, piling up in the dried-out ravine behind the house. Often, the family goes to sleep hungry. In all, 43 per cent of children in Congo suffer from

chronic malnutrition or stunting. The Democratic Republic of the
Congo comes in at 176th of the 189 countries on the 2018 United Nations
Human Development Index.[2]

The United States comes in at thirteenth on this list—but it has
plenty of examples of poverty too. Harvard sociologist Matthew
Desmond spent eighteen months living in low-income housing in
Milwaukee, initially in a trailer park on the mostly white South Side
and later in a rooming house in the black inner city.[3] He followed
people as they tried to avoid homelessness.[4] By and large, his fundamental
story is depressingly similar for each of them, whether they are single
black women with young children (the largest group), a white former
nurse, a married table-dancer, or an ex-soldier who lost both of his legs
to frostbite while coming off a drug high in an unheated house. All of
these individuals are living in rented accommodation, often with children in local schools. Their rental payments account for 70 per cent or
more of their welfare payments or job income. They fall behind with
their rent and are evicted. Their possessions are dumped on the pavement or put into storage—which costs money. They find somewhere
else to live—somewhere that is smaller, in a worse neighbourhood,
where the schools are less good. And the cycle goes on. Being homeless
makes it hard to lead a life that is flourishing.

Poverty matters. Poverty means that a person lacks sufficient
material possessions or income for their needs. What each of us 'needs'
in terms of material possessions or income can be understood in a
number of ways. For a start, it rather seems to depend on where one is
living and at what time in history. In most parts of the world today,
children need shoes but in some parts of the world this is not the case—
as it was not in the past for more places. For example, contemporary
Maasai children generally go barefoot; indeed, Maasai adults typically
spend their days barefoot or wear 'traditional' shoes made of car tyres.[5]

Each of us has basic material needs that follow from our biology.
Someone who is hungry, thirsty, sick, or sleep-deprived is unlikely to
be flourishing. In an attempt to provide a universally agreed definition,
the United Nations in 1995 concluded that 'Absolute poverty is a
condition characterized by severe deprivation of basic human needs,
including food, safe drinking water, sanitation facilities, health, shelter,
education, and information. It depends not only on income but also on
access to services.'[6]

Absolute poverty is also referred to as extreme poverty, abject poverty, destitution, or penury. Although it is easier to live on a few dollars

a day in some countries than in others, a simple definition in terms of income enables changes over time to be determined and comparisons among (and within) countries to be made. A commonly used figure is $1.00 a day in 1996 US prices—equivalent to US$2.20 in 2020.

Until recently in human history, the large majority of people lived under conditions that we nowadays characterize as absolute poverty. They supported a small elite (*plus ça change . . .*). Encouragingly, whatever precise figure for daily income is used, and contrary to what is supposed by many people, both the number and percentage of people living in absolute poverty has gone down considerably over the past two hundred years, especially over the past fifty years (Figure 2.1).

The removal of many people from poverty has various causes. Chief among them are education and the wider use of technologies in such fields as agriculture, medicine, communications, and transport.[7] Against these are factors such as failure to use technology to mitigate the effects of natural disasters, irresponsible exploitation of natural resources, corruption, and war. Violence contributes to poverty; by some measures it has decreased over the past century, despite the Great War and the Second World War, though that is small consolation to those experiencing violence now.[8] However, the rate of progress in

World population living in extreme poverty, 1820–2015
Extreme poverty is defined as living on less than 1.90 international-$ per day.
International-$ are adjusted for price differences between countries and for price changes over time (inflation).

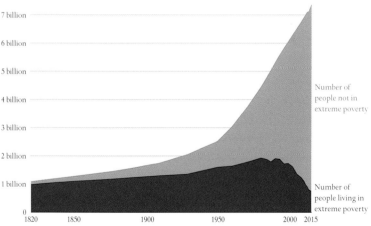

Figure 2.1 The number and the proportion of people living in extreme poverty continues to fall.

tackling extreme poverty is slowing, with the World Bank now thinking that its target of fewer than 3 per cent of people living in extreme
poverty by 2030 will not be reached.[9] Indeed, the number of people in
extreme poverty in many countries in Africa and in the continent as a
whole has increased in recent years.

These rather bald statistics are rendered more immediate and memorable when personalized. In the terrible days of the famines in Ukraine
(the Holodomor), an official was making a speech about the millions of
people dying of hunger. Joseph Stalin interrupted him to say, 'If only
one man dies of hunger, that is a tragedy. If millions die, that's only statistics.'[10] With rather more compassion, Mother Teresa said, 'If I look at
the mass, I will never act. If I look at the one I will.' Alleviation of poverty
needs both scientific insight for efficient deployment of technology and
spiritual wisdom to use it to promote flourishing.

The relationship between human flourishing and the satisfaction of material needs is not straightforward

Some of the great nineteenth century novelists were shrewd observers
of poverty. Think Jean Valjean (who rises from poverty) and Fantine
(who sinks as a result of it) in Victor Hugo's *Les Misérables*; Oliver Twist,
the eponymous character in Dickens' novel; or Tom, the child chimney
sweep, who drowns at the start but is restored to life through his
exercise of virtue in Charles Kingsley's *The Water-Babies*, which is partly a
tract against child labour. But how can we compare material poverty
across very different countries?

Purchasing power parity (PPP) uses the prices of specific goods to
provide a measure for comparing the absolute purchasing power of the
currencies of different countries. Simple versions of PPP have even been
compiled using local prices of internationally available products such
as fast food items or phones and computers. The international dollar is
a conceptual unit of currency with the same PPP as the US dollar has in
the USA. The use of PPP shows that material circumstances are not
straightforwardly related to satisfaction with one's life or even (a narrower issue) with one's standard of living.

Figure 2.2 presents the relationship between the average PPP-adjusted
gross domestic product (GDP) per person in various countries and the

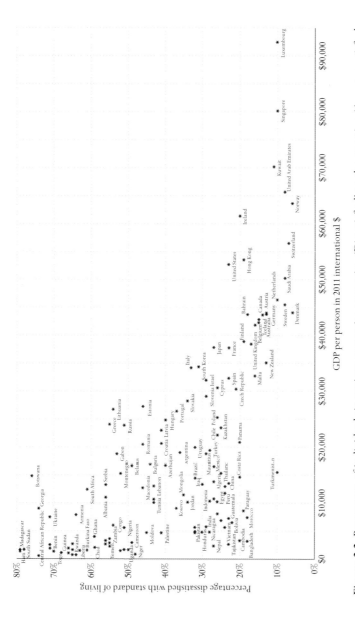

Figure 2.2 Percentage of individuals in various countries answering 'Dissatisfied' to the question 'Are you satisfied or dissatisfied with your standard of living, all the things you can buy and do?'

percentage of the population in each country who report that they are dissatisfied with their standard of living.[11] There are three things of note. First, by and large, up until a value of GDP per person of about international $40,000, there is an inverse relationship between the wealth of a country and the percentage of its population who say that they are dissatisfied with their standard of living—with the figure reaching 75 per cent or more in some of the poorest countries. Second, once a country has a GDP per person of about international $40,000, further wealth brings little further increase in the percentage of population who say they are satisfied with their standard of living. Third, there is a huge amount of scatter, especially in poorer countries—compare Botswana and Bangladesh.

These three features are also found in the relationships between happiness and income within countries. Because different researchers use different measures for wealth and different measure for happiness or satisfaction, it is difficult to make more than rough comparisons across studies. Setting aside the considerable scatter that is found, people are happier the more money they have—but only up to a point. Once they have more than can be bought in the US for about $40,000 a year, more money doesn't on average result in greater happiness. Globally, $40,000 a year is a very large income—far more than the great majority of people have. Other studies suggest that the figure at which more money no longer leads to greater happiness is much lower. In the USA, the average per-person after-tax income in 2009 dollars more than tripled between 1957 and 2014, yet self-reported happiness measures hardly changed across the period. In 1957, 35 per cent of people said they were 'very happy'; in 2014 the figure was 33 per cent.[12]

What other factors apart from income contribute to happiness? The relative importance of different factors varies from country to country but the World Happiness Report for 2017 (yes—such a document exists) finds that GDP per head and the extent of social support (having someone to count on in times of trouble) are the two most important factors, followed by how many years of healthy life the average citizen can expect to live, freedom to make life choices, generosity (recent donations made), and trust (perceived absence of corruption in government and business). These six variables account for three-quarters of the variation between countries.[13]

As is likely to surprise few people, the top five places in the index of world happiness are taken by Norway, Denmark, Iceland, Switzerland,

and Finland, while the bottom five (of the 155 countries for which we have data) are occupied by Central African Republic, Burundi, Tanzania, Syria, and Rwanda. Less expected though, unless one knows something of the recent history of the countries, the greatest increases in happiness from 2005–7 to 2014–16 (of the 126 countries for which we have data for those years) are for Nicaragua, Latvia, Sierra Leone, Ecuador, and Moldova while the five countries with the greatest decreases in happiness are Venezuela, Central African Republic, Greece, Botswana, and Ukraine.

The World Happiness Report is not alone in attempting to find what, in addition to money, makes for happiness. One approach that is increasingly being used refers to 'multidimensional poverty'. A standard definition, deriving from the work of Sabina Alkire and others, states that 'Multidimensional poverty encompasses the various deprivations experienced by poor people in their daily lives—such as poor health, lack of education, inadequate living standards, disempowerment, poor quality of work, the threat of violence, and living in areas that are environmentally hazardous, among others.'[14] Another prosperity index evaluates inclusiveness of society, openness of economy, and empowering of people. In 2020 it was found that health improved in all regions except for North America, and better education and living conditions contributed to improved flourishing everywhere. Denmark came top. In several countries it was possible to attribute greater flourishing to conscious government choices.[15]

How to be happy?

Winning the lottery isn't all it is generally assumed to be. There are many accounts of people who win a lottery who end up saying that they wish they hadn't. When Billie Bob Harrell Jr won the $31 million Texas Lotto jackpot in June 1997 he thought his problems were over.

> Nearly broke and constantly moving between low-paying jobs, with a wife and three children to support, the first of his $1.24 million annual payouts seemed like the light at the end of the tunnel. Instead, it was the beginning of an *annus horribilis* for the 47-year-old Texan. It started out joyful: he quit his job at Home Depot, took his family to Hawaii, donated tens of thousands of dollars to his church, bought cars and houses for friends and family, and even donated 480 turkeys to the poor. But his lavish spending attracted unwanted attention, and he had to change his phone number several times after strangers called to demand donations.

He also made a bad deal with a company that gives lottery winners lump-sum payments in exchange for their annual checks that left him with far less than what he had won. When Harrell and his wife Barbara Jean separated less than a year later, it was the straw that broke the camel's back. His son found him dead inside his home from a self-inflicted gunshot wound on May 22, 1999, shortly before he was set to have dinner with his ex-wife. While family members disputed the idea that Harrell could have committed suicide, he clearly wasn't happy with his life; he'd told a financial adviser shortly before his death that 'Winning the lottery is the worst thing that ever happened to me.'[16]

A more systematic analysis of the consequences of winning a lottery on human happiness was provided by researchers who interviewed people who had won what at the time were large amounts of money (typically, several hundred thousand US dollars) in a lottery. The researchers used standard psychological survey responses to determine their happiness and then compared the results with non-lottery winners from the same geographical area.[17] They found that the lottery winners were no happier and derived significantly *less* pleasure from everyday activities such as hearing a good joke or receiving a compliment.

The same researchers also interviewed people who had become paraplegics or quadriplegics as a result of an injury within the past 12 months. Unsurprisingly, their happiness scores were lower than both the lottery winners and a control group who had no special circumstances or conditions. However, the differences were small and the paraplegics and quadriplegics reported *more* enjoyment of everyday pleasures than the lottery winners. In addition, all three groups reported similar expected future happiness levels.

Locked-in syndrome

Locked-in syndrome is a condition that usually results from a stroke that damages part of the brainstem. Virtually the entire body, including most of the facial muscles, is paralysed but the person is conscious and retains an ability to perform certain eye movements, which may allow them to communicate. The condition is perhaps best known through *The Diving Bell and the Butterfly*, the memoir of the French journalist Jean-Dominique Bauby who developed locked-in syndrome in 1995 as a result of a stroke (Figure 2.3).

Bauby's memoir has sold in its millions in French and in translation. To write the book, Bauby worked with a transcriber who repeatedly

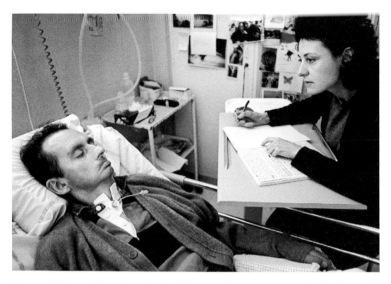

Figure 2.3 Jean-Dominique Bauby 'dictating' his memoir to Claude Mendibil.

recited the letters E S A R I N T U L O M D P C F B V H G J Q Z Y X K W—the order of the frequency with which they occur in French. When the transcriber got to the correct letter, Bauby would blink his left eyelid, the one part of his body he could still control. An average word took about two minutes to produce and the book took ten months of 'dictation', four hours a day. He recounts his experiences of being paralysed, including the poignancy of his feelings on a rare day out from the hospital at the beach with his family. *Le Scaphandre et le Papillon* was published on 9 March 1997. Bauby died of pneumonia two days later.

In one quite large study of what it feels like to live with locked-in syndrome, data were gathered on 65 members of the French Association for LIS (locked-in syndrome).[18] The study and its findings are summarized in the title of the publication that resulted: 'A survey on self-assessed well-being in a cohort of chronic locked-in syndrome patients: happy majority, miserable minority'. Figure 2.4 is a histogram of how people with locked-in syndrome felt about their lives on a scale that range from −5 ('As bad as the worst period in my life'), through 0 ('Neither well not [sic] bad') to +5 ('As well as in the best period prior to LIS').

The majority of the patients said that they were happy with their lives. This result agrees with other research showing that people with

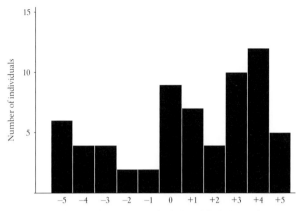

'Self-reported measures by individuals with locked-in syndrome
on the present qualityof their lives: −5 = 'As bad as the worst period in my life';
+5 = 'As well as in the best period prior to LIS''.

Figure 2.4 Self-reported measures by 65 individuals with locked-in syndrome on how they feel about their lives.

severe disabilities may report a good quality of life despite being socially isolated or having major difficulties in activities of daily living, and it is likely to strike those of us who have never experienced anything like locked-in syndrome as remarkable. For our purposes, it suggests at least two things. First, humans often have a great ability to adapt to new circumstances, however disadvantageous they may appear to an outsider. Second, human flourishing is not only not just about how much money one has, it is also not just about one's physical health.

Maslow's hierarchy of needs

Abraham Maslow was an American psychologist at a time when most psychology was about mental abnormalities and illnesses. Maslow wanted to know what constituted positive mental health. He formulated what became known as Maslow's hierarchy of needs (Figure 2.5). Our most fundamental needs are things like food and water, but once these are met, 'higher' needs emerge.

> Obviously a good way to obscure the 'higher' motivations, and to get a lopsided view of human capacities and human nature, is to make the organism extremely and chronically hungry or thirsty. Anyone who attempts to make an emergency picture into a typical one, and who will measure all of man's goals and desires by his behavior during extreme

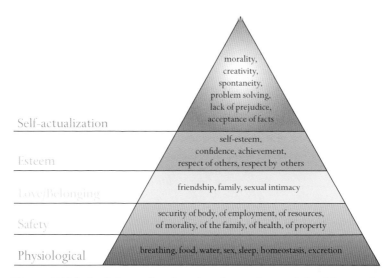

Figure 2.5 Maslow's hierarchy of needs—the needs at each level have to be met before people can spend much effort attending to the needs at higher levels.

> physiological deprivation is certainly being blind to many things. It is quite true that man lives by bread alone—when there is no bread. But what happens to man's desires when there is plenty of bread and when his belly is chronically filled?[19]

Figure 2.5 illustrates how, as each 'layer' of our needs is met, it can support the next layer. Eventually, we have the 'space' to achieve our full potential, to 'self-actualize'. Maslow later introduced another layer at the top of the pyramid: transcendence. As he put it 'Transcendence refers to the very highest and most inclusive or holistic levels of human consciousness, behaving, and relating, as ends rather than means, to oneself, to significant others, to human beings in general, to other species, to nature, and to the cosmos.'[20] Because of its pinnacle contribution to human flourishing in all its fullness, we shall devote the whole of Chapter 4 to transcendence.

What is meant by 'human flourishing'?

It is time to look in more depth at what we said about human flourishing in Chapter 1. We will return at various points in this book to the issue of what is meant by human flourishing, gradually building up a richer picture. There are a number of phrases that capture some of the meaning of 'flourishing', such as 'being happy', 'being positive about life', 'having a sense of well-being' and 'feeling satisfied with one's life'. It is not possible to make clear-cut distinctions between these phrases, but they—and others like them—seem to be used by most people in ways that, while they may overlap, do have certain differences between them. These differences include the time span over which such assessments are made and whether the assessments are subjective or claim to be objective.

'Now' versus 'in one's life to date'

Imagine you are asked, whether face-to-face or as a written survey question, 'How happy do you feel?' The question is principally to do with the here and now. It wouldn't surprise us to find that the same person answers quite differently at different times during the day, depending, for example, on whether they have just had a nice conversation with a colleague, friend, or family member or, conversely, just trapped one of their fingers in a door or realized that they have lost their phone. If the question is changed to 'How happy do you feel now?', those responding are even more likely to give answers that vary over time—for all that any instantaneous account of our happiness is affected by our life to date.

On the other hand, imagine a question like 'Overall, how happy do you consider your life to be?' Here there is more of a requirement for the respondent to reflect on the question rather than answering automatically, and to produce some sort of time-aggregated rather than snapshot response. This is even more the case if the question is 'How happy do you consider your life to have been?', where the expectation that a time-aggregated response will be made is not only explicit but intended to refer to the whole of the respondent's life to date.

Subjectivity versus objectivity

If the time span over which measures of happiness, well-being, or flourishing are one variable that affects such measures, another variable is whether the measurement is required to be subjective or objective. Subjective measures, in contemporary society, are easy to understand. After all, whether *I* am happy or flourishing is surely up to *me* to decide? If I say I am unhappy, who are you to disagree, whatever the evidence you muster to support your assertion?

However, imagine that there do exist objective accounts of happiness and flourishing that are true in all places and for all times. In that case, all our actions, experiences, and feelings need examining *sub specie aeternitatis*, a phrase first used by the Dutch philosopher Baruch Spinoza. It means 'in the light of eternity'. Spinoza (1632–77) worked as an optical lens grinder but was also one of the early thinkers of the Enlightenment. Growing up in a Portuguese-Jewish community in Amsterdam, he rejected the notion of the providential God of Abraham, Isaac, and Jacob, and any claims that the commandments of the Torah and rabbinic legal principles were given by God and binding on Jews. His free thinking got him into trouble and a coalition of rabbis and Calvinist clergy had him expelled from Amsterdam. His most famous work, his *Ethics*,[21] was published after his death. In it, Spinoza tries to deal with ethics as objectively as Euclid dealt with mathematics.

To this day there are still arguments about whether ethics is an objective discipline. If flourishing is subjective, each of us can, indeed, needs to, construct our own path towards it. However, if flourishing is objective, the path is one that we have to *find* rather than *construct* for ourselves. It might be something of each.

Happiness

Much of the empirical literature on human flourishing relies on people's accounts of their self-reported happiness. The term 'happiness' is sufficiently nebulous that some have even argued that it cannot be defined. However, the philosopher Daniel Haybron cautions against such a counsel of despair. Haybron argues that while we may not be able to provide a single account of happiness that captures all the ways in which the term is used, 'we can ask why we *care* about happiness, and whether some proposed definition makes sense of our practical concerns'.[22]

Haybron argues that happiness is best thought of as a *psychological condition*—a word for a certain state of mind. This helps us to distinguish it from a life that goes well for us, which requires a *value* judgement. The way that Haybron draws the distinction is sufficiently clear that it helps to introduce it here:

> When the ancient Greek philosopher Aristotle (384–322 CE) said something about 'happiness'—his word was *eudaimonia*—he was talking about the value notion. And his concern was not to understand a state of mind. He wanted to know what sort of life ultimately benefits a person, serves

her interests or makes her better off. Suppose a man leads a pleasant life of utter passivity, living like a pig and letting his passivity go to waste. Can we really be doing well? This is a question of values, not psychology.[23]

Within the understanding of happiness as a psychological condition, Haybron points out that we can distinguish three basic theories:

- Happiness as pleasure (hedonism);
- Happiness as a positive emotional condition (emotional state theory);
- Happiness as being satisfied with one's life (life satisfaction theory).

The simplest of these three to understand is hedonism. Hedonism argues that only pleasure and its opposite (usually stated as 'suffering') are the components of well-being. The doctrine of hedonism has arisen more or less independently in many cultures and at different times in history. In the Western tradition it is perhaps most centrally associated with Jeremy Bentham in the late eighteenth and early nineteenth centuries and John Stuart Mill in the nineteenth century. Of these two, Bentham had the 'purer' version, arguing that we should be able to calculate units of pleasure—consisting of the product of their intensity and their duration. The morally right course of action is then that which maximizes the excess of pleasure over suffering, aggregated over all beings capable of experiencing pleasures and pains. Bentham is therefore one of the parent figures in the animal welfare movement, famously arguing:

> The French have already discovered that the blackness of the skin is no reason why a human being should be abandoned without redress to the caprice of a tormentor. Perhaps it will some day be recognised that the number of legs, the hairiness of the skin, or the termination of a tail are equally insufficient reasons for abandoning to the same fate a creature that can *feel*. What else could be used to draw the line? Is it the faculty of reason or the possession of language? But a full-grown horse or dog is incomparably more rational and more conversable than an infant of a day or a week or even a month old. Even if that were not so, what differ-ence would that make? The question is not *Can they reason*? or *Can they talk*? but *Can they suffer*?[24]

There are problems with Bentham's account of hedonism, for all that it made a major contribution to thinking about ethics and provided great support to organizations that campaigned against animal cruelty. For a start, it's fair to say that no one has ever managed convincingly to undertake any of the calculations about which Bentham enthused. How does one manage to compare the pleasure of eating with the pleasure of enjoying the company of others or of listening to music? Those who support Bentham argue that such calculations, although fiddly to do, are in principle doable, especially if one favours a particular version of hedonism which takes seriously an individual's preferences, so that the way to compare the pleasure of eating with the pleasure of going to a concert is simply to see which of these alternatives individuals, in particular circumstances, prefer.

A different objection to Bentham's version of hedonism was voiced by the polymath J. S. Mill. In his essay *Utilitarianism*, he wrote 'It is better to be a human being dissatisfied than a pig satisfied; better to be Socrates dissatisfied than a fool satisfied. And if the fool or the pig think otherwise, that is because they only know their own side of the question.'[25] Mill's point is that some pleasures are at a higher 'level' than others. It is fair to say that acres of ink have been spilt by philosophers debating this point. For one thing, it smacks of a cultural elitism to which we are perhaps more sensitive nowadays than in Mill's time. If I enjoy going to the opera, reading Dostoevsky, and drinking Premier Grand Cru Bordeaux on a regular basis (I wish), do I really experience more pleasure from so doing compared to you if you enjoy raves, reading Barbara Cartland, and downing vodka and diet coke? It would be hedonistic snobbery to regard one as superior to the other.

Emotional state theory goes beyond hedonism. Whereas under hedonism to be happy is for one's moment-to-moment experiences to be sufficiently pleasurable, under emotional state theory to be happy is for one's psychological condition to be well disposed towards one's life. Emotional state theory thus takes more account of the fact that happiness is an interaction between who we are and the experiences we have. Most of us know people who, by and large, enjoy what they are doing, whether it is work, time with family and friends, or dealing with the day-to-day issues that we are all faced with. Equally, most of us know people who unfortunately, whether or not they technically suffer from depression, anxiety or any other mainly negative medical conditions, are typically unhappy.

Finally, life satisfaction theory, as its name suggests, is all about any of us looking at our life as a whole and evaluating how satisfied we are

about it. It can entail simply considering one's life at the present
(whether one has a job or not and how one feels about that, how one
feels about one's family, one's friends, one's health, and so on) or it can
entail looking back over the whole of one's life and making a judgement
about its overall quality—something older people are more likely to do
than younger ones. In a study of 1500 people over the age of 65 in the
USA who were asked to look back on their lives, their biggest regrets,
ranked in order, were as follows:[26]

1. Not being careful enough when choosing a life partner;
2. Not resolving a family estrangement;
3. Putting off saying how they felt;
4. Not traveling enough;
5. Spending too much time worrying;
6. Not being honest;
7. Not taking enough career chances;
8. Not taking care of their body.

More positively, there are countless examples of people looking back
over their lives and finding great satisfaction even if much of it was
spent in extremely tough circumstances. Nelson Mandela's *Long Walk to
Freedom* gives his account of his early life, coming of age, and 27 years
imprisoned for his role as a leader of the then-outlawed African
National Congress.[27] Maya Angelou's *I Know Why the Caged Bird Sings*, the
first volume of her seven-volume autobiography, illustrates how her
strength of character and love of literature helped her to overcome
racism and childhood rape.[28] Corrie ten Boom's *The Hiding Place* relates
how she survived Ravensbrück and other concentration camps to
which she and her sister had been sent for helping Jews survive in Nazi-
occupied the Netherlands.[29] Each of these people flourished not because
of their material conditions, which were often terrible, but because of
choices which they made about the purpose which they would pursue
and the meaning which they would seek. That is why we shall devote
Chapters 6 and 7 to those topics.

Life satisfaction theory seems to be able to take into account both the
good and the bad things that happen to us in a less reductionist way
than hedonism does. Most of us, if only occasionally, reflect on how
our day has gone or, at New Year's Eve, on how the year has gone. We
may also think about our close relationships or our jobs in this way,
attempting to produce an overall judgement as to how satisfied we are

with them. It gives scope for scientific insight about the constraints of this material world in which we conduct our lives, and for spiritual wisdom about the values which we adopt in making our choices.

Measuring well-being

It's all very well discussing happiness, life satisfaction or well-being, but can they be validly measured? The Cambridge philosopher of science Anna Alexandrova approaches the issue from a distinctive perspective. Whereas philosophers generally seek for unifying theories—in this case, a single theory of well-being—scientists are more likely to focus on data, in this case, the well-being of specific kinds of people in particular contexts. What Alexandrova does is to bring out the best in both approaches by fusing the empirical concerns of scientists (how things *are*) with the normative considerations of philosophers (how things *should be*).

As an example, think how you might build a theory of child well-being which could be measured. This would be what Alexandrova calls a mid-level theory, intermediate between the high-level abstractions of philosophers and the day-to-day specific of psychologists, medics, economists, and other natural and social scientists. Children do well, Alexandrova maintains, to the extent that they fulfil two conditions, each of which is necessary, and which together are sufficient:[30]

1. Develop those **stage-appropriate** capacities that would, for all we know, equip them for **successful future**, given their **environment**.
2. And engage with the world in **child-appropriate** ways, for instance, with curiosity and exploration, spontaneity, and emotional security.

In Condition 1, 'stage-appropriate' means that children of different ages and abilities are honing different capacities. One expects a typical fifteen year old to have substantially different capacities (e.g. in reading, writing, speaking, and listening) from a five year old, and one expects those who have musical or athletic gifts to have different capacities in these areas from the rest of us. The phrase 'successful future' acknowledges the fact that while child well-being is in the here and now, childhood is also a preparation for adulthood. The 'environment' is important for the straightforward reason that good childhoods vary across time and place—for example, a good childhood in many countries nowadays is helped by the acquisition of various skills that were not so necessary, indeed, may not have existed, in the past (think, knowing the mobile number of one's parent/carer). Condition 2 ensures that childhood well-being is not only about preparation for adulthood (in which case a terminally ill child could not flourish), but also about being a child now—engaging with the world with curiosity and exploration, spontaneity, and emotional security.

One can try to go further in specifying what a good childhood consists of. The National Trust is a large UK charity dedicated to protecting Britain's buildings, landscapes, and coastlines. They have produced an imaginative list of 50 things that children should do before they are eleven and three-quarters: Get to know a tree, roll down a really big hill, camp outdoors, build a den, skim a stone, go welly wandering, fly a kite, spot a fish, eat a picnic in the wild, play conkers, explore on wheels, have fun with sticks, make a mud creation, dam a stream, go on a wintry adventure, wear a wild crown, set up a snail race, create some wild art, play pooh sticks, go paddling, forage for wild food, find some funky fungi, get up for the sunrise, go barefoot, join nature's band, hunt for fossils and bones, go stargazing, climb a huge hill, explore a cave, go on a scavenger hunt, make friends with a bug, float in a boat, go cloud watching, discover wild animal clues, discover what's in a pond, make a home for wildlife, explore the wonders of a rock pool, bring up a butterfly, catch a crab, go on a nature walk at night, help a plant grow, go swimming in the sea, help a wild animal, watch a bird, find your way with a map, clamber over rocks, cook on a camp fire, keep a nature diary, watch the sunset, and take a friend on a nature adventure.[31] Between the two of us we reckon we have done all of these, though not all by the time we were eleven and three-quarters.

To David Lewis, whom we met at the start of this chapter, this would sound like another world—and it is. The National Trust is not concerned in this exercise with producing a definitive account of child well-being but with trying to get more children in Britain to do enjoyable and worthwhile things out-of-doors, things that it seems today's children are less likely to do than we were. The National Trust's list does capture some components of 'valuable child-like ways of being': unstructured, imaginative play…time spent outdoors and in the natural world…open exploration and exhilaration unburdened with previous knowledge.

Measures of well-being cannot be totally objective in the sense of value-free. For example, an Aristotelian who holds that an immoral person cannot flourish might insert some measure of virtue into their assessment of well-being, thus concluding that well-being is very low among sociopaths (to the presumed bafflement of certain sociopaths). Similarly, a psychologist who uses only hedonic measures of well-being would conclude that the proverbial tortured artist (think Vincent van Gogh) has low levels of flourishing however much they reject any other life.

Scientific advances in measuring well-being need to be explicit about their underlying assumptions. The theory of the phenomenon and its measurement *co-evolve*.[32] Rather as people with locked-in syndrome can help those who are not severely disabled to question their presumptions about the implications for well-being of living with disability, so the findings of studies on well-being, once analysed and interpreted, have the potential to feed back into and refine our understanding of the nature of well-being.

Inequalities and well-being

So far in this chapter, there has been almost nothing about the importance of others for the level of human flourishing that any of us enjoys. The main place we address this is in the next chapter, Chapter 3, on the relational dimension of flourishing, the ways in which our relationships with others affects us. But there is another way in which others affect our flourishing and that is by the extent to which they enjoy more or fewer material benefits than we do, in other words, the extent of material inequalities in society.

Ronald Regan was the first American president to make as a mainstay of his policy that 'All Americans have the right to be judged on the sole basis of individual merit and to go just *as far as their dreams and hard work will take them*' (italics added). Bill Clinton adopted the slogan and used it frequently, as did George W. Bush, John McCain, and Marco Rubio. Barack Obama used it more than all previous presidents combined, 'as far as their talents and their work ethic and their dreams can take them.'[33] The slogan lives on. In his moving speech on 7 November 2020, the day that his tally passed the 50 per cent threshold of 270 Electoral College votes, Joe Biden declared, 'I know I've always believed, and many of you heard me say it, I've always believed we can define America in one word:

possibilities. That in America everyone should be given an opportunity to go as far as their dreams and God-given ability will take them.'[34]

Barack Obama, in the words quoted above, was speaking about education. Our (Andrew's and Michael's) universities are passionate about an admissions policy based solely on merit. Time and resources are poured into encouraging applications from schools whose students have not historically gained admission (or even applied) to our institutions, and our colleagues take endless pains to allocate places to those who in their judgement will make the best use of the opportunity.

Let us suppose that American presidents and university admissions processes achieve their intentions 100 per cent. Where does that leave those who are admitted to good universities? We hope that they will make great use of the opportunity, and that they flourish as a result. There is a risk that they will waste the opportunity. There is a risk that the institutions will put more effort into imparting technical academic skills than the values that need to underpin them, perhaps because there is no agreement on what those values should be or the responsibility of educational establishments to equip their students morally. And there is also a risk that individuals will take all the credit themselves, believing that they have succeeded solely because of their own merit. After all, that is what the whole system is telling them.

When we (Andrew and Michael) look back on our own careers, we see a rather different story. We did not choose our parents, neither the genes which they passed on to us nor the physical and spiritual and moral and intellectual nurturing of our early years. We each benefited from teachers who poured their passion for their subject into inspiring and informing us and training our minds to think. As our careers progressed we benefited from professors and others who mentored us and created opportunities for us, and from those who founded and funded the institutions in which we worked. And, dare we say it, we benefitted from luck, so that in the jumble of randomness and non-linearity that make up the world as we encounter it we have somehow been able to do what we have done. We would never belittle individual responsibility—far from it—but we would want our satisfaction in our achievements to go hand in hand with gratitude for those who have equipped us for them.

But now there is a darker side to meritocracy. What about those who don't make it? What about those whose dreams and God-given ability don't seem to take them very far, at least in society's metrology? Insofar as we have achieved a perfect meritocracy, they have only themselves to blame. Few people would advocate reverting from meritocracy to aristocracy, but by itself meritocracy simply provides another way of separating the haves from the have-nots. Except that now it is the have-nots' own fault.

The book that did the most to bring to attention the importance of material inequalities in society for human well-being was *The Spirit Level: Why More Equal Societies Almost Always Do Better*,[35] published in 2009 and authored by two epidemiologists, Richard Wilkinson and Kate Pickett. The same two authors followed it up in 2018 with *The Inner Level: How More Equal Societies Reduce Stress, Restore Sanity and Improve Everyone's Well-being*,[36] and founded The Equality Trust, which works to improve the quality of life in the UK by reducing economic and social inequality. Their work builds and extends that of others who have looked at the consequences of inequality. Indeed, back in 1968, Gary Becker argued that would-be criminals make a cost–benefit assessment of the anticipated rewards of their activities (the benefit—from their perspective) versus the chances of their getting caught and punished (the cost).[37] Accordingly, the greater the material inequality in a society, the more one would expect there to be burglaries and other types of theft.

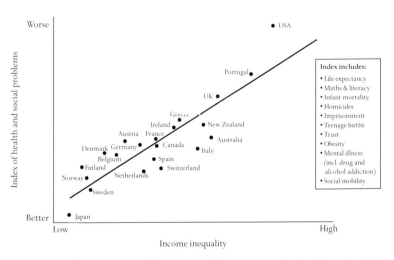

Figure 2.6 The more unequal a country, the worse are its health and social problems.

Figure 2.6 illustrates how the more unequal a country is, the worse are the health and social problems of the people in it. Although the axes lack numbers, they are based on quantitative analysis.[38] The measure of income inequality used is the ratio of how much richer the top 20 per cent of people are compared to the bottom 25 per cent. The index of health and social problems is a combination of data for nine different aspects, equally weighted. This kind of analysis is not without its critics, but it has also been defended by independent experts.[39] Similar patterns are found in countries where data exist for different geographical areas within that country; health and social problems are worse in US states with greater income inequality. Income inequality (and other measures of economic inequality) have risen considerably in many countries in recent decades and continue to do so.[40] The same is true at a global level.[41]

When the analysis is restricted to wealthy countries, income inequality within a country is a better predictor of that country's health and social problems than is the average income of the country. Figure 2.7 illustrates this with regard to child well-being.

What is going on? Why is it that greater income inequality is correlated with a whole basket of factors such as lower life expectancy, poorer school performance, greater infant mortality, more homicides, more imprisonments, more teenage births, lower levels of trust, higher levels of obesity, and higher levels of mental illness?

It seems that the adverse consequences of raised levels of inequality often result from the comparisons that we make with others. These comparisons may be conscious or unconscious. Greater levels of envy, consumerism, isolation, alienation, social estrangement, and anxiety all follow from greater levels of inequality. Across the world, people are becoming more angry, stressed, and worried.[42] While inadequate levels of material provision matter, in many countries what matters more for the quality of people's lives is the inequality within that country. A damaging effect of comparison with others is that it can undermine self-esteem and the respect of an individual for and from others.

Although Figure 2.5 does not show Maslow's later addition of transcendence at the top, it does show how esteem builds on other foundations. Esteem rests on love and belonging, typically manifested through friendship, family, and sexual intimacy. These all lie in the domain of relationships, which is the next of our three dimensions of human flourishing.

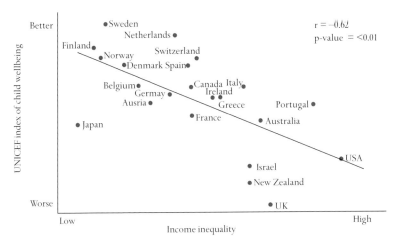

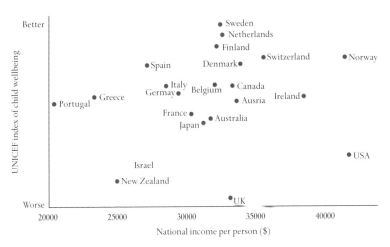

Figure 2.7 In wealthy countries, income inequality within a country is a better predictor of child well-being than is the average income within the country.

Notes

1 Ross, A. (2015) A Congolese child's tale: David Lewis Dieumerci at 10 years old, *The Guardian*, 31 August. Available at https://www.theguardian.com/global-development/2015/aug/31/africa-children-democratic-republic-congo-drc-david-lewis-dieumerci-10-years-old.

2 United Nations (2018) *Human Development Indices and Indicators: 2018 Statistical Update*, United Nations, New York. Available at http://hdr.undp.org/sites/default/files/2018_human_development_statistical_update.pdf.

3 Desmond, M. (2016) *Evicted: Poverty and Profit in the American City*, London: Allen Lane.

4 Markovits, B. (2016) The heartbreaking story of becoming homeless in America, *The Spectator*, 9 April. Available at https://www.spectator.co.uk/2016/04/the-heartbreaking-story-of-becoming-homeless-in-america/.

5 Choi, J. Y., Suh, J. S. and Seo, L. (2014) Salient features of the Maasai foot: Analysis of 1,096 Maasai subjects, *Clinics in Orthopedic Surgery* **6**(4), 410–19.

6 United Nations (1996) *Report of the World Summit for Social Development, Copenhagen, 6–12 March 1995*, United Nations, New York, p. 38. Available at https://undocs.org/pdf?symbol=en/a/conf.166/9.

7 Browne, J. (2019) *Make, Think, Imagine: Engineering the Future of Civilisation*, London: Bloomsbury.

8 Pinker, S. (2011) *The Better Angels of Our Nature: A History of Violence and Humanity*, London: Allen Lane. Pinker's work is controversial. While many have praised it, many have criticized it either on the grounds that some of the data, particularly from prehistoric populations, are 'cherry-picked' or because of Pinker's espousal of atheism and confidence in human reason. Many of the historical data, though, are from mainstream and widely accepted sources.

9 World Bank (2018) *Decline of Global Extreme Poverty Continues but Has Slowed*. Available at https://www.worldbank.org/en/news/press-release/2018/09/19/decline-of-global-extreme-poverty-continues-but-has-slowed-world-bank.

10 Quote Investigator (2019) A single death is a tragedy; a million deaths is a statistic. Available at https://quoteinvestigator.com/2010/05/21/death-statistic/.

11 Roser, M. and Ortiz-Ospina, E. (2017) Global Extreme Poverty, *Our World in Data*. Available at https://ourworldindata.org/extreme-poverty.

12 Myers, D. G. and DeWall, C. N. (2018) *Psychology*, 12th edition, New York: Worth.

13 Helliwell, J., Layard, R. and Sachs, J. (2017) *World Happiness Report 2017*, New York: Sustainable Development Solutions Network. Available at https://s3.amazonaws.com/happiness-report/2017/HR17.pdf.

14 Oxford Poverty and Human Development Initiative (2019) Policy—A multidimensional approach. Available at https://ophi.org.uk/policy/multidimensional-poverty-index/.

15 Legatum Institute (2020) *The Legatum Prosperity Index: A Tool for Transformation*, 14th edition. Available at https://docs.prosperity.com/2916/0568/0539/The_Legatum_Prosperity_Index_2020.pdf.

16 Pous, T. (2012) The tragic stories of the Lottery's unluckiest winners, *Time*, 27 November. Available at http://newsfeed.time.com/2012/11/28/500-million-powerball-jackpot-the-tragic-stories-of-the-lotterys-unluckiest-winners/slide/billie-bob-harrell-jr/.

17 Brickman, P., Coates, D. and Janoff-Bulman, R. (1978) Lottery winners and accident victims: Is happiness relative?, *Journal of Personality and Social Psychology* **36**, 917–27.

18 Bruno, M.-A., Bernheim, J. L., Ledoux, D. et al (2011) A survey on self-assessed well-being in a cohort of chronic locked-in syndrome patients: happy majority, miserable minority, *BMJ Open* **1**, e000039. Available at https://bmjopen.bmj.com/content/bmjopen/1/1/e000039.full.pdf.

19 Maslow, A. H. (1943) A theory of human motivation, *Psychological Review* **50**, 370–96. Available at http://psychclassics.yorku.ca/Maslow/motivation.htm.

20 Maslow, A. H. (1971) *The Farther Reaches of Human Nature*, New York: Penguin, p. 269.

21 Spinoza, B. (1677) *Ethics* in *The Collected Writings of Spinoza*, 2 vols, translated by Edwin Curley, Princeton, NJ: Princeton University Press (vol. 1: 1985).

22 Haybron, D. M. (2013) *Happiness: A Very Short Introduction*, Oxford: Oxford University Press, p. 9.

23 Ibid., p. 11.

24 Bentham, J. (1789) *An Introduction to the Principles of Morals and Legislation*. Available at https://www.earlymoderntexts.com/assets/pdfs/bentham1780.pdf, pp. 143–4.

25 Mill, J. S. (1863) *Utilitarianism*, London: Parker, Son and Bourn. Available at https://www.earlymoderntexts.com/assets/pdfs/mill1863.pdf, p. 7.

26 Pawlowski, A. (2017) How to live life without major regrets: 8 lessons from older Americans, *TODAY*, 17 November. Available at https://www.today.com/health/biggest-regrets-older-people-share-what-they-d-do-differently-t118918.

27 Mandela, N. (1994) *Long Walk to Freedom*, New York: Little, Brown.

28 Angelou, M. (1969) *I Know Why the Caged Bird Sings*, New York: Random House.

29 Ten Boom, C., with Sherrill, J. and Sherrill, E. (1971) *The Hiding Place*, Lincoln, VA: Chosen Books.

30 Alexandrova, A. (2017) *A Philosophy for the Science of Well-Being*, New York: Oxford University Press, p. 69.

31 National Trust (2012) '50 things to do before you're 11¾' activity list. Available at https://www.nationaltrust.org.uk/features/50-things-to-do-before-youre-11--activity-list.

32 Alexandrova (2017) *A Philosophy for the Science of Well-Being*, p. 127.

33 Sandel, M. J. (2020) *The Tyranny of Merit: What's Become of the Common Good?*, London: Alan Lane, pp. 67–8.

34 Biden, J. (2020) Joe Biden and Kamala Harris Election Acceptance and Victory Speech Transcripts November 7. Available at https://www.rev.com/blog/transcripts/joe-biden-kamala-harris-address-nation-after-victory-speech-transcript-november-7, 24:21.

35 Wilkinson, R. and Pickett, K. (2009) *The Spirit Level: Why More Equal Societies Almost Always Do Better*, London: Allen Lane.

36 Wilkinson, R. and Pickett, K. (2018) *The Inner Level: How More Equal Societies Reduce Stress, Restore Sanity and Improve Everyone's Well-being*, London: Allen Lane.

37 Becker, G. S. (1968) Crime and punishment: An economic approach, *Journal of Political Economy* **76**(2), 169–217.

38 The Equality Trust, Notes on statistical sources and methods. Available at https://www.equalitytrust.org.uk/notes-statistical-sources-and-methods. Income inequality is averaged across the reporting years 2003–6 from the United Nations Development Program Human Development Reports, such as that at http://hdr.undp.org/en/content/human-development-report-2006.

39 At the time of writing (February 2021), the Wikipedia entry for *The Spirit Level* provides a good overview of the arguments contradicting and supporting Wilkinson and Pickett: https://en.wikipedia.org/wiki/The_Spirit_Level_(book). Their methods are outlined at https://www.equalitytrust.org.uk/about-us.

40 Goldin, I. and Muggah, R. (2020) *Terra Incognita: 100 Maps to Survive the Next 100 Years*, London: Century, pp. 171–97.

41 Legatum (2018) *2018 Key Findings: Prosperity is Growing, but not Equally*, Dubai: Legatum.

42 Gallup (2019) *Gallup 2019 Global Emotions Report*. Available at https://www.gallup.com/analytics/248906/gallup-global-emotions-report-2019.aspx.

3

The Relational Dimension

Solitary confinement

Thomas Silverstein (Figure 3.1) died on 11 May 2019, aged 67. He had lived in solitary confinement since he killed a prison guard in 1983. He is believed to have spent the longest time in solitary confinement of anyone in the US. Silverstein had originally been sent to prison for a robbery but while in prison he killed two inmates and then the prison guard. In 2008 he wrote to a friend 'It's almost more humane to kill someone immediately than it is to intentionally bury a man alive.'[1]

Those of us who have never been on our own for long, indeed who may wish at times for greater solitude, can find it difficult to realize how soul destroying (to use terminology that might be more appropriate in Chapter 4) it can be to be in solitary confinement. Its negative consequences for mental health have been appreciated for at least two hundred years. A recent review concluded 'There is unequivocal evidence that solitary confinement has a profound impact on health and wellbeing, particularly for those with pre-existing mental health disorders, and that it may also actively cause mental illness.'[2] Solitary confinement can cause a range of health problems including hallucinations (auditory and visual), difficulties with thinking, concentration, and memory, disturbances of thought, problems with impulse control, claustrophobia, rage, severe depression, withdrawal, blunting of affect, and apathy.

Prisoners confined to solitary confinement for more than about ten days often state that it becomes difficult to distinguish between reality and their own fantasy world. As one might expect, symptoms intensify as time goes by. One former US prisoner, who himself had spent two years in solitary confinement, reported:

Figure 3.1 Thomas Silverstein spent 36 years in solitary confinement in the US.

> I have seen inmates lose their mind completely because of the sound of a light where they are yelling at the light, cursing at the light, believing that for some reason the [authorities] planted some kind of noise inside the light purposely . . . and so the inmates that ain't strong minded, don't have something to hang on to, the light, the sound of the door, can make them lose their mind . . . I found it strange, you know, how can a grown man, a very big, grown man, break down to a light. But that's what [that place] can do. And once you lose your mind, you don't know right from wrong. You don't know that you're breaking a rule. You don't know what to do exactly.[3]

The United Nations, in its so-called 'Nelson Mandela' rules, defines prolonged solitary confinement as solitary confinement for more than fifteen days and has called for it to be prohibited (along with such things as the use of constantly dark or lit cells, corporal punishment, the restriction of diet or drink, and collective punishment).[4] It is therefore somewhat chilling to find that 99.9 per cent of the US prison population in solitary confinement—around 80,000 individuals at any one time—spend longer than fifteen days in solitary confinement, with 82 per cent of them spending more than a month, 24 per cent spending longer than a year and 5 per cent spending longer than six years.[5]

Most of the studies of the effects of solitary confinement have been on prisoners. Few of us have been prisoners, but many of us at some point in our life will be lonely.

Loneliness

Loneliness is not the same as being alone. Someone may choose to live on their own and have little contact with others, and yet not experience loneliness (feeling lonely). Many people live on their own yet have lots of social relationships and rarely feel lonely. Someone else, as the charity Mind helpfully puts it, 'may have lots of social contact, or be in a relationship or part of a family, and still feel lonely—especially if [they] don't feel understood or cared for by the people around [them].'[6]

Loneliness can have many causes—for example, experiencing a bereavement, the break-up of an important relationship, retiring, going off to university, moving home, or changing jobs. Certain times of the year can be tough (Christmas is often a problem for some) and factors such as having no close friends or family, being a single parent, belonging to a minority group, or having a disability or long-term health issues make it more likely that a person will be lonely (Figure 3.2). People can be lonely, whatever their age, including children of kindergarten age.

Various scales have been devised to determine whether someone is lonely. Mostly, these are subjective—what someone says about how they are feeling, rather than the extent to which they meet a set of objective criteria such as how often they talk with other people. It may be that loneliness is a single construct or that it has a number of discrete components. One commonly used scale, the de Jong Gierveld (dJG) Loneliness Scale, has separate sub-scales for emotional loneliness and for social loneliness.[7] The emotional loneliness sub-scale is captured by such items as 'I experience a general sense of emptiness' (loneliness is indicated by agreement with the statement); the social loneliness sub-scale by such items as 'There is always someone I can talk to about my

Figure 3.2 Woman with a robotic seal in a Japanese care home. Loneliness can have many causes.

day-to-day problems' (loneliness is indicated by disagreement with the statement).

In the US, Asia, Europe, and a number of other countries various surveys suggest that about one in four adults over the age of 65–70 are lonely.[8] It is difficult to be sure, but a classic study by Robert Putnam argued that we are more solitary nowadays than we used to be.[9] In the US, people have become less likely than used to be the case to join political parties or trade unions, volunteer, or know people in their community. Occasionally one hears of people who have been dead for years before their bodies have been discovered. On 25 January 2006, officials repossessing a bedsit in London found the skeleton of a 38-year-old woman lying on the sofa. The television set was still on and a small pile of unopened Christmas presents lay on the floor. Food in the refrigerator was marked with 2003 expiry dates. The woman was Joyce Carol Vincent.[10] The film maker Carol Morley subsequently made a moving documentary about her: *Dreams of a Life*.

The health consequences of loneliness can be severe.[11] Adults who are lonely are more likely to smoke cigarettes, to sleep badly, eat less well, and take less exercise. An oft-quoted statistic is that the adverse effect of loneliness is equivalent to smoking about 15 cigarettes a day or being clinically obese, though it is difficult to untangle cause and effect.[12] The data on the health consequences of solitary confinement and loneliness indicate that for most of us, it's tough to be on one's own. We are a social species.

Our early relationships

Our initial closest relationship is with the woman within whom we began our post-conception life. Once a baby is born, there can be many individuals who start to play a role in looking after it, though in most cultures it is still the baby's own mother who spends the most time with the newborn and who is most likely to be the principal provider of food, warmth, care, and love.

Other adults and older children can also play important roles in looking after newborns. In Bali, there is an ancient custom that decrees that an infant's feet should not touch the ground until it is 105 days of age. Most people in Bali practise a form of Hinduism in which there is a belief that newborns are still close to the sacred realm from which they came. Newborns therefore deserve to be treated with especial respect and the ground is seen as a source of pollution.[13] A consequence is that, while awake, a baby is held at almost all times by the mother, the father, adult relatives, or neighbours.

Some of the classic work on the importance of relationships in childhood was done by the psychiatrist John Bowlby in England in the 1930s to 1950s.[14] Bowlby observed that children experienced considerable distress when separated from their mothers. Even when other caregivers provided food, the children, bereft of their mothers, were anxious. At that time in England, mothers were generally the main carer. Before

Bowlby and the work of others who also carefully observed mothers and their children, a prevalent understanding of mother–child relationships was informed by a behaviourist perspective, largely resulting from studies in rats, which presumed that the main reward that children got from their mothers was not holding and attention but food and the satisfaction of other material needs.

After he graduated in 1928, Bowlby undertook volunteer work at a school for what were then termed 'maladjusted children' while he decided what to do for a career.[15] He was led to train as a child psychiatrist following his experiences with two boys. One was an isolated, remote, affectionless teenager who had been expelled from his previous school for theft and had had no stable mother figure; the other was anxious and trailed Bowlby around to the extent that he became known as Bowlby's shadow. Subsequent work by Bowlby with John Robertson lead to an influential and upsetting film *A Two-Year-Old Goes to Hospital*.[16] The film documents the emotional damage to a two-year-old girl, Laura, following an eight-day spell in hospital for a minor operation. As was typical at the time, parental visits were severely restricted. Her mother was absent and there was little continuity in Laura's nurses. At first, Laura was upset at her mother's absence but then, ominously, she became quiet. When her mother appeared again after the eight days, Laura's trust in her mother had been deeply shaken and she was withdrawn in her behaviour. Other studies showed that such parental absence could have very long-lasting and severe consequences for children.

A Two-Year-Old Goes to Hospital highlights the central importance of relationships for human flourishing. Visiting arrangements in children's wards have improved greatly since the film was made.

Transitional objects

In the early 1950s, the British psychoanalyst Donal Winnicott noticed that between about four and twelve months after birth, an infant often starts to attribute a special value to something soft, like a piece of cloth or a toy, such as a teddy bear (Figure 3.3). Winnicott talked about such an object being the first 'not-me possession'.[17] In common with other psychoanalysts, Winnicott held that the newborn baby is not yet aware that it and its mother are not one. In many ways, this is hardly

Figure 3.3 Many young children have a comfort blanket or something similar, like a toy, that serves as a transitional object.

surprising—until birth they were, in a sense, one and, more mundanely, the newborn baby lacks the mental capacity to realize that it is not all there is in the world.

As a child begins to become aware that its mother is a separate entity, it feels it has lost something—especially as such awareness of the distinction between itself and its mother is likely to come when it is hungry and there is a delay in its hunger being satisfied. This awareness results in frustration and anxiety. The transitional object can help at this time. To the growing child, such an object can stand in for the mother and allow the child to fantasize that she is still available. A transitional object can be so important to a young child that attempts by a parent to remove the object, even if only briefly to wash it, can result in a distraught, even traumatized, child. It is for this reason that parenting guides nowadays frequently advise that you get your child when young used to two such objects—though this can be easier said than done.

Eventually, if all goes well in development, a child is able to hold in its mind a representation of its mother or other significant primary caregiver and the transitional object ceases to be so important. However, if the child is rushed, forced to give up its transitional object too soon, there is a danger that its development is harmed. Winnicott saw playing with a transitional object as an important stage in the child beginning to understand how it relates to others and to the world in general.[18] During this stage, it would be a mistake for an adult to try to explain to the child that its transitional object is make believe or that the child's imaginary friend does not exist. Indeed, the creativity the child displays in its play helps it better both understand reality and develop its own sense of self.

An evolutionary perspective on the importance of human relationships

There are many species where the parents effectively abandon their offspring early in life—think of all those tadpoles that frogs produce and then leave as they hop away. Humans don't normally do that. As a species we produce relatively few offspring and parents can invest in them heavily. Cultures vary in the age at which offspring are considered ready to leave their parents but somewhere between 12 and 20 years is typical. Before then there is nine months of pregnancy, anything up to a couple of years of breast- or bottle-feeding, and then years of childhood and adolescence. Much of this time is spent learning the skills that are essential to leading a life away from the parental home.

But we don't just learn from our parents. The psychologist Nick Humphrey worked at the University of Cambridge Sub-Department of Animal Behaviour at Madingley, a village some three miles from Cambridge, at a time when the animal behaviourist Robert Hinde headed up the group. In a famous essay, Humphrey argued for how important other individuals can be in mammalian cognitive development:

> in working with laboratory monkeys I have been mindful of the possible damage that may have been done to them by their impoverished living conditions. I have looked anxiously through the wire mesh of the cages at Madingley, not only at my own monkeys but at Robert Hinde's. Now, Hinde's monkeys are rather better-off than mine. They live in social groups of eight or nine animals in relatively large cages. But these cages are almost empty of objects, there is nothing to manipulate, nothing to explore; once a day the concrete floor is hosed down, food pellets are thrown in and that is about it. So I looked—and seeing this barren environment, thought of the stultifying effect it must have on the monkey's intellect. And then one day I looked again and saw a half-weaned infant pestering its mother, two adolescents engaged in a mock battle, an old male grooming a female whilst another female tried to sidle up to him, and I suddenly saw the scene with new eyes: forget about the absence of objects, these monkeys had each other to manipulate and to explore. There could be no risk of their dying an intellectual death when the social environment provided such obvious opportunity for participating in a running dialectical debate. Compared to the solitary existence of my own monkeys, the set-up in Hinde's social groups came close to resembling a simian School of Athens.[19]

What is true for monkeys is even more true for us. Our social relationships are crucial for our development when we are young and for our quality of life when we are older. Extreme evidence is provided by historical cases of badly run orphanages and by rare instances of feral children.

A feral child is someone who from an early age lived with little or no meaningful contact with other humans. Stories in fiction include Romulus and Remus, the twins in Roman mythology who were suckled by a wolf before being brought up by a shepherd and his wife, and Mowgli, also raised by wolves, in Kipling's *The Jungle Book*. The stories in real life end badly.

A well-documented, and horrifying, case is that of Genie.[20] She looked about six or seven—it turned out she was 13—when she hobbled with her mother into a Los Angeles county welfare office in October 1970. Ever since she had been a toddler she had been strapped into a home-made straitjacket and tied to a chair or potty in her home in Temple City, California, by her deranged father, who later killed himself. He had forbidden her to cry, speak, or make a noise and had beaten and growled at her, like a dog.

Initially, Genie could only speak a few words, including just two phrases—'stop it' and 'no more'—and mostly remained silent and undemonstrative. She moved with a sort of bunny hop and was doubly incontinent, urinating and defecating when stressed. Doctors described her as the most profoundly damaged child they had ever seen. Progress was initially encouraging. She learned to play, dress herself, and enjoy music. She expanded her vocabulary and sketched pictures. However, forming words into sentences proved beyond her, bolstering the view that beyond a certain age it is simply too late to develop this ability. Humans seem in our development to have 'sensitive periods' for a number of things. If we miss these sensitive periods, subsequent development of the skill in question is much harder and may not even be possible. There were arguments and legal cases about who should look after Genie; by the time she had reached her late twenties she had, if anything, gone backwards in her development.

The overthrow in 1989 of the Ceausescu regime in Romania led to a widespread awareness that thousands of children were being kept in inadequate state orphanages in the country. Aside from poor material circumstances, perhaps the main problem with such orphanages was the lack of adult stimulation for the children. Nurses often lacked much training and had to cope with large numbers of children. Studies were subsequently undertaken on children taken from these orphanages and rehomed. In one study conducted three years after children had been adopted, it was found that the longer children had spent in such orphanages, the less secure their attachment to their new parents (as Bowlby would no doubt have predicted), the greater their behaviour problems, and the lower their intelligence scores.[21] They also scored higher for 'indiscriminate friendliness'—in which children behave much the same towards any adults as they do towards their caregivers (not a good strategy in evolution, given that other adults may be less likely to have a child's best interests at heart).

Getting off heroin

A rather different illustration of the importance of human relationships comes from a phenomenon that greatly surprised health officials in the US back in the early 1970s. At that time, the US had spent 15 rather fruitless years in the Vietnam War. War is profoundly unsettling to many combatants and research revealed that about 35 per cent of US soldiers in Vietnam had tried heroin and about half of these had become addicted. In other words, 15–20 per cent of US service personnel in the Vietnam War were addicted to heroin.[22]

Lee Robins was one of the researchers charged with following up addicted service members once they returned home. In a finding that completely overturned accepted beliefs about drug addiction, Robins found that when soldiers who had been heroin users returned to the US, only 5 per cent of them were addicted a year later, with just 12 per cent becoming readdicted within three years of their return.[23]

What was going on? The prevailing wisdom at the time was that it is extremely difficult to get off heron, once one is addicted to it. Indeed, as a rule of thumb, 90 per cent of those who come out of heroin rehab subsequently relapse—yet the Vietnam data showed remarkably different rates of relapse. In Vietnam the soldiers had been in the presence of other addicted, bored, and stressed soldiers. Once back in the US, they were in a relatively heroin-free environment, often with family and long-standing friends. Their relationships had changed, and so did their behaviour.

Dunbar's number

Robin Dunbar is a distinguished anthropologist with a background in both evolutionary biology and psychology. He pointed out that the more people we know, the more difficult it becomes to maintain social relationships with them. He then observed that most of us can only comfortably maintain stable relationships with about 150 individuals. This figure has become known as Dunbar's number. He describes it as 'the number of people you would not feel embarrassed about joining uninvited for a drink if you happened to bump into them in a bar'.[24] Although most famous for the number 150, Dunbar has argued that we devote about two-thirds of our social time to just 15 people, and within that number there is an inner core of about 5 individuals who take up about 40 per cent of our social time.

These numbers give us a handle on the different layers of social intimacy that most of us enjoy. We may have over a thousand names in our e-address books but chances are there are quite a few of these we have completely forgotten about—some may even have died. We can maintain reasonable ties with about 150 people and then there are progressively smaller numbers who matter progressively more to us. For many of us, once we have left our natal family home, there may be one person who is especially important to us.

And the Lord God said, It is not good that the man should be alone

In the language of the King James' Bible, one of the early Hebrew accounts of the creation of humans reads 'And the Lord God said, It is not good that the man should be alone.'[25] Plenty of adults, of course, live contentedly on their own. But the majority of people live for much of their adult life in a sexual relationship with, at any one time, one other person, even if some are not always sexually faithful.

In the animal world, there are a range of ways in which males and females pair up to produce offspring—and there are some asexual species too, meaning that they don't have males and that adult females manage quite satisfactorily on their own. Humans, on the basis of such things as the extent of our sexual dimorphism and genetic analyses of variation on our X and Y chromosomes, can be predicted to be what animal behaviourists would characterize as 'mildly polygynous'. In truly monogamous species, an adult male only mates with one adult female and vice versa. If one of these two dies, the remaining individual may be able to find a new mate with which to pair—an instance of serial monogamy. In mildly polygynous species, some males mate (over a short period of time) with more than one female, some males mate with just one female and other males never get to mate.

Analyses of our sex chromosomes and the genes in our mitochondria (which we inherit only from our mothers) indicate that in the past humans were more polygynous than we are in many societies now.[26] Most countries prohibit bigamy—in which one person can be married to more than one person at the same time. Polyandry (one woman married to more than one man) is much rarer than polygamy (one man married to more than one woman).[27]

Most adults marry, although in about 80 per cent of countries marriage rates fell in the 35 years from 1970 to 2005.[28] In part this fall corresponded to increasing rates of cohabitation (i.e. living together without being married). Whatever the explanations for such trends, it remains the case that married people are generally happier than unmarried people (Figure 3.4), whether compared with people who have never been married, people who are separated, or people who are divorced.

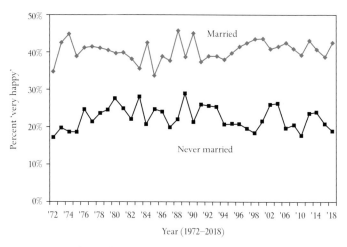

Figure 3.4 Married people are more likely than unmarried people to report that they are very happy. Data from the USA.

Not only are they happier, married people are healthier too, though this is more marked for men than for women. A possible explanation for this, which both the male authors of this book find pretty convincing, is that quite a few men are not much good at looking after themselves. Not every health consequence of marriage is positive—married people are more likely to be overweight, for example.[29] Yet, by and large marriage is beneficial, always remembering that these are average results, so there will be exceptions, and that the data are correlational so that it might be that happier and healthier individuals are more likely to get married.[30]

Nevertheless, despite occasional protestations in the popular press to the contrary, it really does seem that being married is usually good for one's health. Perhaps the most convincing data come from longitudinal studies in which people are followed for many years during which researchers note both their health (using standard objective measures) and their marital status.[31] Such studies show that people's health improves after they get married, compared to people who do not get married. There may be a number of reasons for the health benefits of marriage but they seem to be to do with living with someone rather than with the legal or religious status of marriage. Cohabitation, at least in the US, has very similar health benefits to marriage,[32] which is quite a notable finding given that cohabiting relationships are typically less stable (i.e. less persistent) than married ones.

Marriage is found in all societies, though some religious branches or sects either encourage or require their members to be celibate. In Judaism and Christianity, marriage is seen as a relationship in which the two become one. Historically, divorce was either frowned upon or forbidden. Indeed, the strict Roman Catholic position is that one cannot divorce—hence the growth of annulments, in which it is decreed that a valid marriage never took place. In the New Testament, marriage is likened to the relationship between Christ and the Church, so that a marriage is seen as having the potential to reveal a relational aspect of God's love.

Helping behaviour in humans

Helping is at the heart of most positive relationships. A lot of ink has been spilled in not very useful discussions about what precisely is meant by altruism and cooperation and whether a clear-cut distinction can be made between them. A key point is to do with whether, in helping another individual, there is a cost to me. If I am a woodlouse threatened with desiccation and I and another woodlouse cling together to reduce the rate at which we lose water, there is no cost to our behaviour (unless we complicate matters by considering that such behaviour might increase the transmission of disease or lead to competition between the woodlice for some resource such as food) and each of us benefits. But if I am a song bird and I give an alarm call on seeing a predator to warn others in my group, this calls for an explanation—surely, I would have been better off keeping quiet and letting another bird raise the alarm, thereby increasing *its* risk of being spotted and attacked by the predator.

Most of us combine elements of selfishness and helpfulness in our natures. There has been a long-running debate among biologists as to what is the natural state of affairs with regards to these and to why humans are neither more selfish nor more helpful than they are. We are used to thinking how natural selection can be invoked to understand the evolved *structure* of organisms. But Charles Darwin realized that natural selection does not apply only to structures; it applies also to *behaviours*. Darwin further appreciated that the same arguments that apply to the behaviour of non-human animals also apply to humans. His *The Descent of Man, and Selection in Relation to Sex* explored the implications of natural selection for human behaviours.[33] Even though Darwin knew nothing of what today we call 'genetics', he realized that natural selection might still be responsible for the evolution of worker sterility in the social insects. At first sight, such sterility deals a crushing blow for the theory of natural selection, in which evolution takes place by virtue of more successful individuals producing more offspring that survive and in turn contribute to future generations. Workers in such social insects as ants, bees, and wasps produce no offspring—so how can this be functional?

Darwin argued that sterility might evolve by a process he termed 'family selection', nowadays generally known as 'kin selection'. He observed that 'breeders of cattle wish the flesh and fat to be well marbled together; the animal has been slaughtered, but the breeder goes with confidence to the same family'.[34] In other words, both artificial and natural selection do not have to rely on individuals having their own offspring; individuals can reproduce vicariously, as it were, via their close relatives. This can allow altruism—even extreme altruism in which individuals do not reproduce—to evolve and perpetuate.

Darwin's thoughts about altruism largely lay dormant for a century until a PhD student called William D. Hamilton produced a more general, mathematical theory that encapsulated and extended Darwin's insights about the origins of altruism. Hamilton was able to make mathematical predictions about how the extent of altruism should depend on the degree of relatedness between individuals. Advances came thick and fast and the 1960s and 1970s saw an explosion in field work and in theoretical modelling in the disciplines that came to be known as behavioural ecology and sociobiology.[35] Today, there are still areas of disagreement—notoriously with regards to the levels at which selection operates, namely whether selection at the level of genes and individuals is all that needs be considered or whether selection operating between groups of individuals results in phenomena that cannot be explained solely by selection at lower levels.[36] Nevertheless, it is clear that there are two key drivers of helping behaviour in the natural world: kin selection (Darwin's 'family selection') and reciprocal altruism, in which animal A helps animal B and subsequently animal B helps animal A.

Kin selection and reciprocal altruism can be observed in humans. The great majority of people are concerned about the welfare of close relatives (including their children) and those with whom they regularly interact (enabling reciprocal altruism). We humans share much of our biology with our close evolutionary relatives and have been subject over time to the forces of natural selection. It is always risky to attempt to characterize ways in which humans are unique (there is vast research on the extent to which tool use, language, and intelligence are found in other animals), but a defining characteristic of our species is the extent to which we can choose how to behave.[37]

Trolley problems were first devised by the Oxford philosopher Philippa Foot.[38] Trolley problems pose a moral dilemma and then allow us to consider what we feel we should do (our emotional response) and why (our rational response). The classic example—and there are numerous variants—is illustrated in Figure 3.5. You are out on your own one day and see an out-of-control trolley or tram hurtling towards five people who just happen to be tied on tracks ahead of it (this is a thought experiment). There is no doubt that these five will be killed unless—and this is the only alternative to their dying—you pull a lever before the trolley reaches a junction, thus diverting the trolley onto an alternative set of tracks. However, in so doing, you will be

Figure 3.5 Trolley problems probe our moral inclinations and the reasons we give for our intended actions when faced with ethical dilemmas.

directly responsible for the death of one person tied on this alternative set of tracks. Had you not pulled the lever, this person would have survived. Would you pull the lever? Should you pull the lever?

Trolley problems provide the opportunity, being thought experiments, for us straightforwardly to change some of the variables and see what difference that makes. For example, in the above problem it turns out that most people believe that the right thing to do is to pull the lever, killing one person to save five. Nothing, perhaps, very surprising there, agonizing though such a decision might be in real life. But suppose, in a second problem, that you are on a bridge looking down on a single track (no junction) with the trolley hurtling towards the same five individuals and that the only way to save the five is to push someone next to you off the bridge so that (s)he falls onto the track, causing the trolley to derail, but is killed in the process. It turns out that most people say they would not do this. But is there a relevant moral distinction between the two problems?

Books have been written on trolley problems. They have even been applied to the moral dilemmas faced by autonomous vehicles.[39] We won't consider them any more except to note that they illustrate how issues to do with helping behaviour in humans go far beyond what we see in other species. Our cognitive capacities and the natural inclination that most of us have to feel that we need to be able to defend what we do—especially when this harms others—mean that our behaviours, when they have moral implications, are a blend of spontaneous, unthought through actions—as when we rush to help someone who has fallen in the street—and actions that result from consideration—as in trolley problems or, more practically, when we try to decide whether to give money to a particular charity.[40]

To complicate matters further, humans have probably evolved so that at times we can deceive ourselves about the intentions behind our behaviours. The Nobel Laureate Richard Feynman observed that in science, 'The first principle is that you must not fool yourself—and you are the easiest person to fool.'[41] Another Richard, Richard Alexander, pointed out that self-deception enables us the better to deceive others (the actor who believes their performance is real is the more convincing).[42] If I convince myself that you are a cold and ungrateful so-and-so, it makes it easier for me to decide not to repay you (to reciprocate) for some benefit you previously gave me.

Nevertheless, the great majority of us do have a sense of obligation.[43] Indeed, the observation that some people don't helps illustrate how pursuing happiness is not an unquestionable goal. Narcissists, who hold unrealistic superior views of themselves, are over-confident and exhibit little shame, guilt, or empathy, tend to be happier than other people[44]— but spread misery in their wake. Fortunately, only a minority of people are narcissists—though narcissists are more likely than the rest of us to end up in positions of authority (they would, wouldn't they?).[45]

When it comes to what most people think are morally appropriate ways of behaving, for all that there are local differences, there are considerable commonalities too. Analysis of large databases on sixty societies across the world showed that helping kin, helping your group, reciprocating, being brave, deferring to superiors, sharing disputed resources and respecting prior possession are considered morally good in all of them.[46] People often (though not always!) behave in ways that we find morally commendable.

In one large experiment, researchers turned in (to banks, hotels, museums, post offices, etc.) a total of 17,303 'lost' wallets in 355 cities in 40 countries, simply stating that they had found them and had to dash off—leaving the person to whom they handed the wallet to deal with the situation. The wallets varied in the amount of money they contained (either no money or US$13.45 in local currency) and each wallet contained the contact details of the 'owner' (one of the research team) so that the researchers could determine whether the person to whom the wallet had been turned in kept it or contacted the 'owner' to return it.[47]

There were very considerable differences between countries—we recommend you lose your wallet or purse in Switzerland, Norway, the Netherlands, Denmark, or Sweden, where around 70–80 per cent of the wallets were returned, not China, Morocco, Peru, Kazakhstan, or Kenya, where the corresponding figure was 10–20 per cent. A remarkable finding was that, contrary to predictions by both members of the general public and professional economists, the wallets with money were more likely to be returned. Across the whole experiment, 40 per cent of the empty wallets were returned compared to 51 per cent of those with US$13.45. In three countries (Poland, the UK, and the US), wallets with the equivalent of US$94.15 were also turned in. In these three countries, 46 per cent of the empty wallets were returned compared to 61 per cent of those with US$13.45 and 72 per cent of wallets containing US$94.15.

In a sub-study of the same research, people said that failing to return a wallet would feel more like stealing when the wallet contained a modest amount of money than when it contained no money, and that such behaviour would feel even more like stealing when the wallet contained a substantial amount of money. Contrary to the glaring exceptions we read about in the newspapers, it seems that the more that is at stake, the more humans are capable of wanting others to flourish. We shall return to that when we consider models of human motivation in Chapter 8.

Relationships with non-humans

So far, this chapter has primarily concerned itself with relationships among people and the contribution such relationships make to human flourishing. In this final section, we look at relationships with non-humans, specifically with inanimate objects, with (non-human) animals and with God.

Pygmalion goes AI

Relationships with human substitutes have long been explored in fiction. Ovid wrote of the Cypriot sculptor Pygmalion who carved a woman out of ivory and then fell in love with her. The story ends surprisingly well—on kissing her, he found that Venus had granted his heart's desire and turned the statue into a woman. They married and had a daughter. The myth was widely taken up in nineteenth- and twentieth-century novels and plays, one of which formed the basis of the musical *My Fair Lady*. In 1996 William Gibson published his novel *Idoru*, in which the rock star Rez wants to marry Idoru, a digital construct who exists only in virtual reality. In 2013 the movie *Her* was released, in which the central character, Theodore Twombly, purchases an operating system upgrade that includes a virtual assistant which uses machine learning to adapt and evolve. Through a wreckage of failed marriages Theodore names the operating system Samantha and falls in love with 'Her'. It does not end well. In 2019 *Machines Like Me* by Ian McEwan was published, in which the robot Adam forms a romantic eternal triangle with its owners. That does not end well either.

Fiction is becoming reality. In Japan, *otaku* are people (commonly thought of as nerds or geeks) who are obsessed with video games and anime. An increasing number of *otaku* say they are finding fulfilment in virtual relationships and have given up on the idea of real-world romance.[48] Akihiko Kondo dated Miku, a cartoon character, for ten years before he married her in a ceremony in 2018 that he regards as a wedding. There were 39 guests and Miku was present as a cuddly toy, wearing a white, lace dress, and a long veil. Kondo is not alone. In 2018, Gatebox, the company that made Kondo a hologram of Miku, started issuing unofficial 'marriage certificates' to customers; 3,700 people took them up on the offer.

This phenomenon is not that far removed from what most of us would call sex dolls, though the academic term 'agalmatophilia' (from the Greek for loving a statue—back to Pygmalion) sounds more respectable. For all that it may seem implausible, or simply too tacky to be credible, an increasing number of men declare, with apparent sincerity, that they love their realistic sex dolls (sex robots are the next stage[49]). Nor should it be assumed that such feelings will blow over; some men have been in monogamous relationships with their sex dolls for over a decade—and today's sex dolls/robots are a lot more realistic and interactive than they were ten years ago. While some of us might expect such a relationship to lack depth or much mutuality, sex dolls/robots do offer what some consider to be advantages, such as not being unfaithful, arguing, or becoming physically less attractive as they age.

Pets

Pets can play a central place in the life of people, whatever their age. Many people think of their pets as members of their family;[50] their deaths can cause deep grief—and not just for children. Acknowledging, once again, that many of the studies are correlational, so that inferences about causation should be made only with caution, having a pet seems to have a wide range of health benefits. For example, people with pets have lower resting heart rates and lower blood pressures.[51] While one reason for this is that keeping pets (notably dogs and horses) may mean that one gets more exercise, even the act of stroking a pet can help lower blood pressure.

A study where people were asked about their pets and the meaning that they have for their owners showed that pets 'augment social interactions in groups, fostering generalized reciprocity and encouraging social trust'.[52] Pets can play an important role in counteracting loneliness, both by virtue of being companions in themselves and because they may lead their owners to have more interactions with others. Many people feel more comfortable talking to a stranger if the stranger has a dog.[53] Of course, it depends on the size and friendliness of the dog!

Pets can be particularly important for only children and for children in one-parent families.[54] A pet can help a child with no siblings develop greater empathy and higher self-esteem. Such children typically become more trusting, feel safer, and have a greater sense of self-confidence; they participate more in social and physical activities and feel more positive about their community.

God

Pets are tangible. For those with a religious faith, their relationship with God is less tangible but can be of central importance. Evangelical and Pentecostal Christians are often likely to describe their relationship with God as being core to their lives and to believe that God speaks to them personally.[55] Worship (Figure 3.6), a regular prayer life, studying the scriptures, and fellowship with other believers can all strengthen this relationship.

Spiritual writings down the ages have documented how hard it can sometimes be to sustain a close relationship with God. In *The Dark Night of the Soul*, penned in the sixteenth century, St John of the Cross argues

Figure 3.6 Worship can strengthen a believer's relationship with God.

that such nights are a necessary purgation on the path to union with God. In her twenty-first century anthropological study of a conservative, vibrant evangelical church in London, Anna Strhan shows how faith in God is both a comfort and a struggle.[56] A complement of faith is doubt and the religious life can be plagued by guilt as the believer feels they have fallen short in their relationship with God. At an inexcusable level, sexual abuse by a member of the clergy is associated with lower levels of spirituality and with reduced likelihood of describing one's relationship with God positively.[57] In selecting these negative experiences, in some cases resulting from culpable crimes, we should not let our filter dismiss the vast catalogue of positive experiences of God. In Chapter 9 we look at the relationship between religious faith and health; in the large majority of cases, the relationship is a positive one.

To those who live in an academic world where reason and ideas are hallowed above all, it can come as something as a shock to be told by the Professor of Writing at Goldsmiths, University of London, that experience comes before assent to propositions. 'It is the feelings that are primary. I assent to the ideas because I have the feelings; I don't have the feelings because I've assented to the ideas.'[58] Spufford adds, as the very last sentence of the book, 'I don't need to point out that I am not any kind of spokesman for the Church of England, do I?'

For many Christians their relationship with God centres on the person of Jesus Christ. Sometimes this follows extended evidence gathering and analysis. One scientist described his experience as a medical student in his twenties as a transition from belief to commitment:

> I had to make a choice. A full year had passed since I decided to believe in God, and now I was being called to account. On a beautiful fall day, as I was hiking in the Cascade Mountains during my first trip west of Mississippi, the majesty and beauty of God's creation overwhelmed my resistance. As I rounded a corner and saw a beautiful and unexpected frozen waterfall, hundreds of feet high, I knew the search was over. The next morning, I knelt in the dewy grass as the sun rose and surrendered to Jesus Christ.[59]

Francis Collins went on to become Director of the Human Genome Project, the first government-funded consortium to sequence the entire human genome, Director of the National Institutes of Health (NIH), with responsibility for a $40 billion annual budget at the time of the coronavirus pandemic, and the 2020 Templeton Laureate, one of the largest prizes awarded annually to an individual. His relationship with Jesus Christ remains foundational not only to his scientific leadership of NIH, but also to his passion that the benefits of progress in treating and preventing COVID-19 should prioritize the poor and the disadvantaged.[60]

We have already touched on the Christian understanding of marriage and the way in which the relationship between a couple in marriage is seen as reflecting Christ's relationship with the Church. The notion of humans as being made in 'the image of God' (*Imago Dei*) is a prominent concept in Judaism, Christianity, and Sufism. In the Hebrew bible it is introduced in the very first chapter: 'Let us make human beings in our image, after our likeness, to have dominion over the fish in the sea, the birds of the air, the cattle, all wild animals on land, and everything that creeps on the earth.'[61]

Precisely what is meant by 'the image of God' has been much discussed.[62] One approach has been to recognize some special quality that humans possess—whether our rationality, our spiritual awareness, or our moral sensibility. If it is in the nature of God to reach beyond Godself and to engage in love with the Universe which God has created, then humans in God's image might be expected to share that capacity. Another, not necessarily contradictory, approach has been to see it as enabling us to enjoy a special relationship with God, a relationship that in the Judaeo-Christian tradition is seen as having been damaged in the Fall but amenable to restoration. In the culture of the early readers of Genesis, the function of an image or likeness was to *represent* someone. A king, for example, would often set up his image in a remote province, which he might not visit in person, to represent himself and his prestige and authority. On this basis, 'to fill the vocation of representing God as his image and likeness, humankind is called to reach *beyond the self in love and communion* with God and the other, which includes other human beings'.[63] These concepts come together in the distinctively human capacity for responsibility.[64]

All this leads into the third of our three dimensions of human flourishing—the transcendent—to which we now turn in Chapter 4.

Notes

1 Almasy, S. and Holcombe, M. (2019) A federal prisoner believed to have spent the longest time in solitary confinement has died, *CNN US*, 23 May. Available at https://edition.cnn.com/2019/05/23/us/colorado-thomas-silverstein-dies/index.html.

2 Shalev, S., Lloyd, M. and Benyon, J. (2008) The health effects of solitary confinement, in S. Shalev (Ed.), *A Sourcebook on Solitary Confinement*, London: Mannheim Centre for Criminology London School of Economics and Political Science, pp. 9–24. Available at http://www.solitaryconfinement.org/sourcebook. Quotation from p. 10.

3 Shalev, S. (2013) *Supermax: Controlling Risk through Solitary Confinement*. Milton: Willan, p. 192.

4 United Nations Office on Drugs and Crime (2015) *The United Nations Standard Minimum Rules for the Treatment of Prisoners (the Nelson Mandela Rules)*. Vienna: United Nations Available at https://www.unodc.org/documents/justice-and-prison-reform/GA-RESOLUTION/E_ebook.pdf.

5 Skibba, R. (2018) The hidden damage of solitary confinement, *Knowable*, 22 June. Available at https://www.knowablemagazine.org/article/society/2018/hidden-damage-solitary-confinement.

6 Mind (2019) *Loneliness*. Available at https://www.mind.org.uk/media-a/3124/loneliness-2019.pdf.

7 Ong, A. D., Uchino, B. N. and Wethington, E. (2016) Loneliness and health in older adults: A mini-review and synthesis, *Gerontology* **62**, 443–9.

8 Ibid.

9 Putnam, R. D. (2000) *Bowling Alone*. New York: Simon & Schuster.

10 Morley, C. (2011) Joyce Carol Vincent: How could this young woman lie dead and undiscovered for almost three years?, *The Observer*, 8 October. Available at https://www.theguardian.com/film/2011/oct/09/joyce-vincent-death-mystery-documentary.

11 Holt-Lunstad, J., Smith, T. B., Baker, M., Tyler Harris, T. and Stephenson, D. (2015) Loneliness and social isolation as risk factors for mortality: A meta-analytic review, *Perspectives on Psychological Science* **10**(2), 227–37.

12 E.g. Campaign to End Loneliness. Risk to health. Available at https://www.campaigntoendloneliness.org/threat-to-health/.

13 Howe, L. (2005) *The Changing World of Bali: Religion, Society and Tourism*. London: Routledge. Rousseau, B. (2017) In Bali, babies are believed too holy to touch the Earth, *New York Times*, 18 February. Available at https://www.nytimes.com/2017/02/18/world/asia/bali-indonesia-babies-nyambutin.html.

14 Bowlby, J. (1958) The nature of the child's tie to his mother, *International Journal of Psycho-Analysis* **39**, 1–23. Bowlby, J. (1988). *A Secure Base: Parent–Child Attachment and Healthy Human Development*. New York: Basic Books.

15 Bretherton, I. (1992) The origins of attachment theory: John Bowlby and Mary Ainsworth, *Developmental Psychology* **28**, 759–75.

16 Robertson, J. (1952) *A Two-Year-Old Goes to Hospital*. Film available for purchase from http://www.robertsonfilms.info/2_year_old.htm.

17 Winnicott, D. W. (1953) Transitional objects and transitional phenomena: A study of the first not-me possession, *International Journal of Psycho-Analysis* **34**, 89–97.

18 Winnicott, D. W. (1971) *Playing and Reality*. London: Tavistock.

19 Humphrey, N. (1976) The social function of intellect, in P. P. G. Bateson and R. A. Hinde (Eds), *Growing Points in Ethology*, Cambridge, UK: Cambridge University Press, pp. 303–17. Available at http://www.humphrey.org.uk/papers/1976SocialFunction.pdf.

20 Carroll, R. (2016) Starved, tortured, forgotten: Genie, the feral child who left a mark on researchers, *The Guardian*, 14 July. Available at https://www.theguardian.com/society/2016/jul/14/genie-feral-child-los-angeles-researchers. See also https://en.wikipedia.org/wiki/Genie_(feral_child).

21 Chisholm, K. (1998) A three-year follow-up of attachment and indiscriminate friendliness in children adopted from Romanian orphanages, *Child Development* **69**(4), 1092–106.

22 Clear, J. (2018) *Atomic Habits: An Easy & Proven Way to Build Good Habits & Break Bad Ones*, New York: Avery.

23 Robins, L. N., Davis, D. H. and Nurco, D. N. (1974) How permanent was Vietnam drug addiction?, *American Journal of Public Health* **64**(Suppl. 12), 38–43.

24 Dunbar, R. (1996) *Grooming, Gossip and the Evolution of Language*, London: Faber and Faber.

25 Genesis 2:18.

26 Dupanloup, I., Pereira, L., Bertorelle, G., Calafell, F., Prata, M. J., Amorim, A. and Barbujani, G. (2003) A recent shift from polygyny to monogamy in humans is suggested by the analysis of worldwide Y-chromosome diversity, *Journal of Molecular Evolution* **57**(1), 85–97.

27 Barash, D. P. (2016) Polygamy in humans, in T. Shackelford and V. Weekes-Shackelford (Eds), *Encyclopedia of Evolutionary Psychological Science*, Cham: Springer. Available at https://link.springer.com/referenceworkentry/10.1007/978-3-319-16999-6_117-1 (behind paywall).

28 Zagorsky, J. L. (2016) Why are fewer people getting married?, *The Conversation*, 1 June. Available at https://theconversation.com/why-are-fewer-people-getting-married-60301.

29 Lipowicz, A, Gronkiewicz, S. and Malina, R. M. (2002) Body mass index, overweight and obesity in married and never married men and women in Poland, *American Journal of Human Biology* **14**(4), 468–75.

30 Stutzer, A. and Frey, B. S. (2006) Does marriage make people happy, or do happy people get married?, *Journal of Socio-Economics* **35**, 326–47.

31 VanderWeele, T. J. (2017) On the promotion of human flourishing, *Proceedings of the National Academy of Sciences* **114**(31), 8148–56.

32 Musick, K. and Bumpass, L. (2012) Reexamining the case for marriage: Union formation and changes in well-being, *Journal of Marriage and Family* **74**, 1–18.

33 Darwin, C. (1871) *The Descent of Man, and Selection in Relation to Sex*, London: John Murray.

34 Darwin, C. (1859) *On the Origin of Species by Means of Natural Selection, or the Preservation of Favoured Races in the Struggle for Life*, London: John Murray, p. 358.

35 Wilson, E. O. (1975) *Sociobiology: The New Synthesis*, Cambridge, MA: Harvard University Press.

36 Dawkins, R. (1976) *The Selfish Gene*, Oxford: Oxford University Press. Sober, E. and Wilson, D. S. (1998) *Unto Others: The Evolution and Psychology of Unselfish Behavior*, Cambridge, MA: Harvard University Press. Nowak, M. A. and Coakley, S. (Eds) (2013) *Evolution, Games, and God: The Principle of Cooperation*, Cambridge, MA: Harvard University Press.

37 Reiss, M. J. (2019) Science, religion and ethics: The Boyle Lecture 2019, *Zygon* **54**(3), 793–807. Briggs, G. A. D., Halvorson, H. and Steane, A. M. (2018) *It Keeps Me Seeking: The Invitation from Science, Philosophy and Religion*, Oxford: Oxford University Press, Chapter 5.

38 Foot, P. (1967) The problem of abortion and the doctrine of the double effect, *Oxford Review* **5**, 5–15.

39 Frank, D., Chrysochou, P., Mitkidis, P. and Ariely, D. (2019) Human decision-making biases in the moral dilemmas of autonomous vehicles, *Scientific Reports* **9**, 13080.

40 Kahneman, D. (2011) *Thinking, Fast and Slow*, New York: Farrar, Straus and Giroux.

41 Leighton, R. and Feynman, R. (1985) *Surely You're Joking, Mr Feynman! Adventures of a Curious Character*, New York: W. W. Norton, p. 343.

42 Alexander, R. D. (1979) *Darwinism and Human Affairs*, Seattle, WA: University of Washington Press.

43 Tomasello, M. (2020) The moral psychology of obligation, *Behavioral and Brain Sciences* **43**, E56. Available at https://www.cambridge.org/core/journals/behavioral-and-brain-sciences/article/moral-psychology-of-obligation/320 01D22714B9F8ED00D9F1AB2BF254D.

44 Papageorgiou, K. A., Gianniou, F. M., Wilson, P., Moneta, G. B., Bilello, D. and Clough, P. J. (2019) The bright side of dark: Exploring the positive effect of narcissism on perceived stress through mental toughness, *Personality and Individual Differences* **139**, 116–24.

45 Braun S. (2017) Leader narcissism and outcomes in organizations: A review at multiple levels of analysis and implications for future research, *Frontiers in Psychology* **8**, 773.

46 Curry, O. S., Mullins, D. A. and Whitehouse, H. (2019) Is it good to cooperate? Testing the theory of morality-as-cooperation in 60 societies, *Current Anthropology* **60**(1), 47–69.

47 Cohn, A., Maréchal, M.A., Tannenbaum, D. and Zünd, C. L. (2019) Civic honesty around the globe, *Science* **365**(6448), 70–3.

48 Hegarty, S. (2019) Why I 'married' a cartoon character, BBC News, 16 August. Available at https://www.bbc.com/news/stories-49343280.

49 Levy, D. (2009) *Love & Sex with Robots*, London: Gerald Duckworth.

50 Casciotti, D. and Zuckerman, D. (2019) The benefits of pets for human health. National Center for Human Health. Available at http://www.center4research.org/benefits-pets-human-health/.

51 Hodgson, K., Barton, L., Darling, M., Antao, V., Kim, F. A. and Monavvari, A. (2015) Pets' impact on your patients' health: Leveraging benefits and mitigating risk, *Journal of the American Board of Family Medicine* **28**(4), 526–34. Available at https://www.jabfm.org/content/jabfp/28/4/526.full.pdf.

52 Ibid., p. 527.

53 Guéguen, N. and Ciccotti, S. (2008) Domestic dogs as facilitators in social interaction: An evaluation of helping and courtship behaviors, *Anthrozoös* **21**(4), 339–49.

54 Hodgson et al. (2015) Pets' impact on your patients' health.

55 Luhrmann, T. M. (2012) *When God Talks Back: Understanding the American Evangelical Relationship with God*, New York: Vintage.

56 Strhan. A. (2015) *Aliens and Strangers? The Struggle for Coherence in the Everyday Lives of Evangelicals*, Oxford: Oxford University Press.

57 McLaughlin, B. R. (1994) Devastated spirituality: The impact of clergy sexual abuse on the survivor's relationship with god and the church, *Sexual Addiction & Compulsivity* **1**(2), 145–8.

58 Spufford, F. (2012) *Unapologetic*, London: Faber and Faber, p. 19

59 Colllins, F. S. (2007) *The Language of God: A Scientist Presents Evidence for Belief*, New York: Pocket Books, p. 225.

60 Collins, F. S. (2020) Acceptance address by Dr. Francis S. Collins: 'In praise of harmony', Templeton Prize. Available at https://www.templetonprize. org/laureate-sub/address-by-dr-francis-s-collins/.

61 Genesis 1:26.

62 Alexander, D. (2017) *Genes, Determinism and God*, Cambridge, UK: Cambridge University Press, Chapter 12.

63 Thiselton, A. C. (2015) The image and the likeness of God: A theological approach, in M. Jeeves (Ed.), *The Emergence of Personhood: A Quantum Leap?*, Grand Rapids, MI: Eerdmans, pp. 184–201.

64 Briggs, Halvorson and Steane (2018) *It Keeps Me Seeking*.

4

The Transcendent Dimension

And I have felt a presence that disturbs me

William Wordsworth's poem 'Tintern Abbey'—or 'Lines Written (or Composed) a Few Miles above Tintern Abbey, on Revisiting the Banks of the Wye during a Tour, July 13, 1798', to give it its full title—was composed after he had been on a walking tour with his sister Dorothy, to whom the poem is addressed. The poem is 159 lines in length. Perhaps its most famous section (lines 93–7) is:

> And I have felt
> A presence that disturbs me with the joy
> Of elevated thoughts; a sense sublime
> Of something far more deeply interfused,
> Whose dwelling is the light of setting suns,

Wordsworth captures here what many of us have felt at some point. That there are times when we feel transported beyond the everyday, when we feel a burst of joy or a sense that there is great meaning to life, that everything is connected or that, somehow, all is well.[1]

Transcendent literally means 'going beyond the ordinary limits'—the notion that within the ordinary we can get a glimpse of that which is more than ordinary. As the artist and poet William Blake put it:

> To see a World in a Grain of Sand
> And a Heaven in a Wild Flower
> Hold Infinity in the palm of your hand
> And Eternity in an hour
>
> ('Auguries of Innocence')

For both Wordsworth and Blake, this sense of the transcendent was a path to God. But we will not make this transition so easily. We live in a world where there is a more sceptical outlook on the claims of religion.

To what extent are 'elevated thoughts' and a feeling that we can hold 'Eternity in an hour' open to everyone?

Individual experiences of the transcendent often seem as though they were tailor-made for that person. They are not like experiencing hot and cold, which are the same for everyone with normal heat sensors in their skin. Describing them can sometimes seem more like trying to tell a colour-blind person what 'red' means. You could explain to them Maxwell's equations, and then define 'red' in terms of a wavelength range. But even many sighted people might struggle to follow your exposition, and you would still have failed to convey what it means to experience red. We hesitate to describe anyone as 'transcendent-blind', because, although we cannot prove it, we suspect that most people are actually rather sensitive to the transcendent dimension of human flourishing.

Some would describe their experience of the transcendent in terms of their relationship with God. So let us lay out some ground work for this. As a spoiler alert, we (the authors) are not going to rest weight on proofs that God exists. We think that rational arguments have value in justifying the reasonableness of belief in God, and in showing that it stands up robustly to the most ferocious intellectual onslaughts. But in our observation it is rare for individuals to be persuaded to change their beliefs on the basis of rational arguments *alone*, though there are plenty of thoughtful people for whom rational arguments have been *influential* in changing their beliefs.[2]

Self-transcendence

The brilliant but idiosyncratic late-nineteenth-century German philosopher, Friedrich Nietzsche, made a brave attempt to explore a kind of transcendence which did not involve a sense that there is something beyond oneself that is greater than oneself. Self-transcendence has been defined as 'that characteristic movement of man in which he continually surpasses himself, in all that he is, all he wishes and all that he has . . . it is that constant tendency possessed by man to always go beyond the already acquired.'[3]

Nietzsche was an avowed atheist. There was for a time in Cambridge some graffiti of his most famous assertion, ' "God is dead"—Nietzsche'. Before long, underneath it some wag added, ' "Nietzsche is dead"— God'! For Nietzsche there was no possibility of transcendence being associated with our awareness of a divine being. The only hope is to drag ourselves up by our bootstraps, as it were. Self-transcendence is therefore all about self-perfection. To be human is to be capable of creating an authentic existence for ourselves. In his *Thus Spoke Zarathustra*, Nietzsche presents (through the character Zarathustra) the idea of the *Übermensch* (Superman) as a goal for humanity—where *über* can also be translated as 'above' or 'beyond' and has connotations of transcendence or superiority.[4]

In the second half of the twentieth century, Nietzsche's *Übermensch* fell out of favour owing to its frequent use by Hitler and other National Socialists in their vision of a biologically superior Aryan master race. *Übermensch* implies *untermenschen* (inferior beings)—a term which in its English original comes not from Nietzsche but from a 1922 book, *The Revolt against Civilization: The Menace of the Under-man*, authored by a Ku Klux Klan member—Lothrop Stoddard. In Nietzsche's work *Übermensch* serves as more of a hope that humans can rise above our pettiness and become what we are capable of. But as the use and abuse of his thinking was to show, something more than a self-contained aspiration is needed.

Ecstasy

If one enters 'ecstasy' into a search engine, the first hit is likely to be the common name for the recreational drug MDMA (3,4-methylenedioxy-methamphetamine). Although some users report anxiety, panic attacks, confusion, and even paranoia, most people who use MDMA describe feeling very happy and energized, with a great upswelling of love for those around them.[5] It is this that gives it its street name, though the positive effects soon wear off and are often followed by feelings of tiredness and depression. MDMA is an amphetamine derivative with some of the pharmacological properties of mescaline.[6] Mescaline is itself a drug immortalized in such accounts as Aldous Huxley's *The Doors of Perception* and Hunter Thompson's *Fear and Loathing in Las Vegas*, each of which, in very different ways, tells how mescaline can shape one's awareness of reality.

The word 'ecstasy' is much older than MDMA. Literally it means 'standing outside one's self' (from classical Greek). An ecstatic experience is therefore one that takes one out of or beyond one's self. From a psychological vantage point, ecstasy is therefore about losing control but feeling great in so doing—a feeling of intense euphoria.

Activities that can cause such a feeling include sexual intimacy, music, and religion. Time may seem to stand still so that one is lost in the moment. For Trekkies,[7] the Nexus was a pathway to ecstasy. In her attempt to explain the Nexus to Captain Picard, Guinan described it as like 'being inside joy'.[8] Inside the Nexus, reality is determined by one's deepest desires.

Sexual ecstasy is particularly associated with orgasm. With music, ecstasy seems to be particularly likely in group dancing. In religion, it can occur both when someone is solitary or with others. Despite its widespread occurrence, some are suspicious of ecstasy, seeing it as irrational, even as evidence of mental illness.[9] Nowadays, though, we know something of its underlying physiological causes. Without in any way wanting to imply that in all situations ecstasy can be reduced to the actions of a particular hormone, endorphins do play a role.

Endorphins ('endorphin' is a contraction of 'endogenous morphine') comprise a group of hormones whose principal function is to inhibit the communication of pain signals. They are produced and stored in the pituitary gland and act in both the central and peripheral nervous systems. In the central nervous system they work by causing more

Figure 4.1 Ecstatic dancing, fuelled by endorphins, has a long history from prehistoric times to today's raves.

Figure 4.2 Whirling dervishes in Istanbul.

dopamine to be produced and released, dopamine being a neurotrans-mitter associated with pleasure. Endorphins are now known to be produced by a wide range of activities including vigorous aerobic exer-cise (the runner's high), sexual intercourse, listening to music, dancing, worship, and eating certain foods including chocolates. These activities differ in the type and intensity of pleasure that they induce.

Today's raves (e.g. Figure 4.1) are probably not so different from cer-tain types of dancing that have gone on since time immemorial. So-called ecstatic dancing is characterized by a sense of abandonment and an absence of formality. In classical Greek mythology, the Maenads (lit-erally 'raving ones') were intoxicated female worshippers of Dionysus, the Greek god of wine. Sufi whirling is a type of physically active medita-tion among Muslims that is still practised by Sufi Dervishes (Figure 4.2).

It has been argued that religion may originally have evolved in an early immersive version with no formal structure and based on trance dancing.[10] Only later did doctrinal religion evolve, probably when humans began to live in larger communities rather than in the smaller groups characteristic of gather-hunter communities. But the evolu-tionary origins of religion cannot be used as an argument against its validity. The evolutionary origin of our capacity for mathematics might help us to understand the fascination of mathematics, but it does not prove or disprove the validity of mathematics. The almost ubiquitous emergence of belief in God in widely disparate cultures does not prove one way or the other whether God exists, but it might prompt us to take the quest seriously.

What can be proved about God?

There was a time when logical proofs for the existence of God were popular. Perhaps the best known is St Anselm's ontological argument of the eleventh century. Anselm begins with the notion of a being than which no greater can be conceived. Anselm argued that if such a being does *not* exist, then a greater being—namely, a being (a) than which no greater can be conceived and (b) which exists—can be imagined. But this is absurd: nothing can be greater than a being than which no greater can be conceived. We therefore have what is sometimes called *reductio ad absurdum* (the Latin for 'reduction to absurdity'). This is a form of argument sometimes used in mathematics in which it is shown that an initial assumption leads to an absurd conclusion (typically, some sort of logical contradiction). Because of this, the initial assumption is rejected. An early Greek example was the proof that the square root of two cannot be a rational number. Anselm claimed to show that the assumption that a being than which no greater can be conceived does not exist is absurd, i.e. impossible. We therefore conclude that there is a being than which no greater can be conceived; such a being is God.

Both the initial assumptions of Anselm's argument and the subsequent reasoning have been questioned. The closeness of such an argument for the existence of God to mathematical reasoning was highlighted by the philosopher Descartes in the seventeenth century. He argued that there is no less contradiction in conceiving of a supremely perfect being who does not exist than there is in conceiving of a triangle whose interior angles do not sum to 180°.[11] Some of the world's most distinguished philosophers have continued to argue about ontological arguments—arguments about the nature of 'being'. Bertrand Russell (1872–1970) was a British philosopher, logician, essayist, and social critic best known for his work in mathematical logic and analytic philosophy.[12] Since both of us have degrees from Cambridge, we are relieved that he was willing to change his thinking as he learned more. 'Against my will, in the course of my travels, the belief that everything worth knowing was known at Cambridge gradually wore off.'[13] Russell noted that it is much easier to conclude that ontological arguments are unconvincing than to say exactly what is wrong with them.

Arguments for the existence of God have a rather mixed history. This may be because humans have a habit of first making up their minds, and then using their considerable powers of rationality to persuade others of their position. It may also be because to prove anything requires something more absolute than what is being proved to which to refer the proof, and what could be more absolute than God? Perhaps we can take a lesson from the kind of Euclidean geometry which is still taught in schools, but in a way different from Descartes. Euclidean geometry takes as a starting point a small number of axioms which cannot themselves be proved, but which provide a basis on which to build a substantial structure of geometry that works and is useful. The analogy must not be pressed too far, because in 1854 the German mathematician Bernhard Riemann gave a lecture which showed how an infinite family of geometries could be constructed which did not depend on Euclid's postulates; that is not our subject here. Rather, it is that it is acceptable to have a basic belief which cannot itself be proved, and then to build on that belief. It can be argued that belief in God is such a basic belief. One can then ask whether such a belief is rational, reasonable, justifiable, or in some other way warranted, which is different from asking whether it can be proved. At its simplest, a belief could be warranted if it is held by thoughtful people in the context of sensible and reasonable behaviour.[14]

Professor Sir Brian Pippard FRS, who was the Cavendish Professor when we were undertaking our PhDs, acknowledged that the observation that many of his colleagues believed in God was an empirical fact. In a letter to *The Times* dated 14 December 1981, which appeared under the headline 'Accounts of the origin of matter', he wrote:

> Having become aware of ourselves and our surroundings, may we not later, in the course of evolving further complexities, have suddenly acquired another sense, the ability to apprehend and even occasionally to make tenuous contact with the source of all things? . . . I cannot, with so many devout believers among my friends, deny the truth of their experience of God simply because I have not shared it myself.[15]

Pippard counted among his friends some of the best minds at the time in Cambridge and beyond. What kind of experience of God might he have been alluding to?

We are aware of many different kinds of experience of the transcendent. We shall survey some of these, aware that our accounts may
meet with very different kinds of response. Some of them may seem so
weird that they simply evoke scepticism, but we include them out of
respect for those to whom they are so real. Others seem to occur only
to a few individuals, and sometimes rarely even to them; we include
those because as in astronomy rare events may tell us something significant. Not all of them presuppose belief in God; indeed, some of
them even deny it.

Has science proved that there is nothing beyond the material world?

The assertion that what can be studied by science constitutes the only
reality is sufficiently well established that it has a name, *scientism*.
Scientism is a junior cousin of *reductionism*, which in its metaphysical
form asserts that all phenomena can be reduced to their underlying
physical causes. Thus, one of the discoverers of the double helix as the
structure of DNA can write:

> 'You', your joys and your sorrows, your memories and ambitions, your
> sense of personal identity and free will, are in fact no more than the
> behavior of a vast assembly of nerve cells and their associated molecules.[16]

This does not work very well as reductionism, since it stops at a particular level of explanation with which Crick was familiar, namely
nerve cells and molecules (though he did also know rather a lot
about DNA). Why not atoms, or protons, neutrons and electrons?
For that matter, why not quarks and gluons and neutrinos and Higgs
bosons? The answer is that different levels of explanation are appropriate for different questions. There are some questions for which
nerve cells and their associated molecules are indeed relevant. But
not all. If you want to tell your true love why you think they are
wonderful, we don't recommend a discourse on nerve cells and their
associated molecules.

Scientism goes further than reductionism and asserts that the methods
of science are the only way of arriving at truth and values. There is
an obvious self-contradiction involved in such an assertion, since it cannot be established scientifically. But beyond that, it has an unsatisfactory air about it. Passionate as the two of us are about science, it is almost

impossible to conceive how anyone could actually live their lives on the basis only of what can be determined through the methods of science. It is to be hoped that human values are not inconsistent with scientific knowledge, and take account of the best of scientific insight, but it is hard to see how the methods of science alone can provide all the wisdom that is needed for human flourishing.

The Harvard University psychologist and philosopher William James, older brother of novelist Henry James, wrote a widely read book with the evocative title *The Varieties of Religious Experience*.[17] In the book, which was based on his Gifford lectures in Edinburgh in 1901–2, he methodologically eschewed religious institutions and concentrated rather on the psychology of religious feelings. Drawing on vast amounts of literature, mainly autobiographical, as well as his own experience, he distinguished positive and negative characteristics.[18] A healthy-minded religious person has a deep sense of the goodness of life, seen with a sky-blue tint. By contrast, a sick soul finds that from the bottom of every fountain of pleasure something bitter rises up, often with an appalling convincingness. Each of these dispositions can arise seemingly involuntarily or through resolute determination.

James finds that religious experience is useful; it may even be a crucial human biological function. That does not make it true, but his own conclusion is that religious experiences connect us with a kind of reality not accessible to what we might call a scientific study of the material world:

> The further limits of our being plunge, it seems to me, into an altogether other dimension of existence from the sensible and merely 'understandable' world.[19]

A myth has become popular that science and religion have invariably been in conflict. Apparently, there are even colleagues in Oxford who still assert this. Historical scholarship is increasingly showing the myth to be untenable. The distinguished historian of science Peter Harrison invites his readers to imagine that a colleague tells you with great excitement that he has found a contemporary account of a war in 1600 between the countries of Israel and Egypt.[20] How should you react? The answer is with scepticism, because in 1600 those land masses did not have those national identities; there were no countries called Israel and Egypt at that time. For most of history there have been no defined subjects called religion and science, at least not with anything much like their present meanings. There is increasing consensus among scholars that the conflict myth was a backstory created in the second half of the nineteenth century by senior American academics such as John William Draper, Professor of Chemistry at New York University, and Andrew Dickenson White, President of Cornell University. Their motives for doing this can be debated, but they were indubitably selective in their choice of facts and their manner of presentation. The persistence of the conflict myth in the face of the evidence is remarkable, but robust scholarship is progressively replacing it with more nuanced and more positive accounts of the great contributions to what we now call science made by men and women of faith.[21]

The Religious Experience Research Unit

One of the more thorough attempts to collate accounts of religious experience was undertaken by a distinguished marine biologist who, like Sir Brian Pippard, was also a Fellow of the Royal Society. Sir Alister Hardy FRS, FRSE, FLS was Linacre Professor of Zoology at the University of Oxford. Much of his academic work was on marine ecosystems. He invented the 'Continuous Plankton Recorder' which, as its name suggests, enables ships to collect plankton without having to stop.

However, Hardy is principally remembered for two other things. First, he proposed what subsequently became known as 'the aquatic ape hypothesis' (expanded on and much popularized by Elaine Morgan[22]), arguing that a number of our anatomical features—such as little body hair and the presence of subcutaneous fat—could be explained by the theory that we spent a crucial phase of our evolutionary past living in shallow seas, by the shoreline.[23] Secondly, and only some time after he retired at the age of 65, he founded in 1969 the Religious Experience Research Unit at Oxford.

Hardy's Centre (which moved to Lampeter in Wales in 2000, Hardy having died in 1985) houses an archive with over 6,000 accounts of first-hand experiences of people from across the world who had a spiritual or religious experience.[24] Here is one such example that happened to a boy:

It was a hot summer Sunday afternoon, I was lying on my back in a copse lost in reverie. I was not really thinking of anything, and then my mind went a blank—suddenly I found myself surrounded, embraced, by a white light, which seemed to both come from within me and from without, a very bright light but quite unlike any ordinary physical light. I was filled by an overwhelming sense of Love, of warmth, peace and joy—a Love far, far greater than any human love could be—utterly accepting, giving, compassionate total Love. I seemed to sense a presence, but did not see anybody . . . I had the feeling of being 'one' with everything, of total unity with all things, and 'knowing' everything—whatever I wanted to know, I 'knew', instantly and directly. And I had the sense of this being utter Reality, the real Real, far more 'real' and vivid than the ordinary every day 'reality' of the physical world.

I do not know how long this lasted, a minute? Back again in this world, I felt thunderstruck. What was that? What did it mean? I felt that it was of great importance and must have some tremendous meaning . . . but what? What was I supposed to do? Why me? Was I being 'called' for anything? I remember at the time being puzzled that the experience did not seem to relate to the 'religion' I was being taught and in which I had been brought up: I saw none of the iconography of Christianity I remember asking myself. Was that God?—but surely not: 'God' wouldn't come to me, an insignificant small boy! But, whilst puzzling over this (and feeling intense chagrin that I was quite unable to remember anything of the wonderful 'knowledge' that I had then enjoyed) I was convinced, beyond all shadow of doubt, of the 'reality' of the experience, the Reality, the overwhelming Love, the 'Oneness' of all things—and this has lasted, despite all reasoning, later 'reductionism', and suggestions that this was just my 'imagination', or that I was 'dreaming', or 'hallucinating'. But, at the time, I could not 'ground' the experience, and I felt that I could not talk to anybody about it, so I locked it away, pondering over it—a very big, unexplained, question.[25]

This account has a number of features in common with many others collected by Hardy: (i) the individual did not speak of it at the time, but the experience made a life-changing impression on them; (ii) it did not 'fit' with what the person had been told of religion but nevertheless seemed spiritual or religious in some way; (iii) despite being hard to reconcile with anything else in the person's life, it was believed as reality.

Hardy gathered such accounts by inviting people to get in touch with him and asking them the question: 'Have you ever been aware of or influenced by a presence or power, whether you call it God or not, which is different from your everyday self?' Here is a second account, this one an experience of an adult:

In 1956 at the age of 23 my husband and I were walking the cliff path from St Ives in Cornwall to Zennot [sic]. It was a bright sunny day in September, bright but not a garish mid-summer sun. My husband was walking his usual forty yards ahead and disappeared over the prow of an incline, so to all intents and purposes I was entirely alone. Although there was no mist the light seemed suddenly white and diffused and I experienced the most incredible sense of oneness and at the same time 'knew what it was all about' it being existence. Of course, seconds later I hadn't the faintest idea what it was all about. However it struck me that the oneness was in part explained by the sensation that the air and space and light was somehow tangible, one could almost grasp it, so that there was not a space which stopped because my human form was there but that my form was merely a continuity of the apparently solid space.

The experience was unbelievably beautiful, and I will never forget the quality of that bright white light. It was awesome.[26]

Each of these two accounts relates events that took place out of doors. Nature, whether animate (Figure 4.3) or inanimate (Figure 4.4), can engender such experiences. Music can also be a trigger, as in this account:

I was sitting one evening, listening to a Brahms symphony. My eyes were closed and I must have been completely relaxed for I became aware of a

Figure 4.3 Nature can sometimes move us deeply.

Figure 4.4 Many of us find photographs of the night sky to be awesome. This image was taken by the Hubble Space Telescope and shows a group of interacting galaxies called Arp, 273,300 million light years from us in the Andromeda constellation.

feeling of 'expansion', I seemed to be beyond the boundary of my physical self. Then an intense feeling of 'light' and 'love' uplifted and enfolded me. It was so wonderful and gave me such an emotional release that tears streamed down my cheeks. For several days I seemed to bathe in its glow and when it subsided I was free from my fears.

I didn't feel happy about the world situation but seemed to see it from a different angle. So with my personal sorrow. I can truly say that it changed my life and the subsequent years have brought no dimming of the experience.[27]

And here is one more, chosen to illustrate the fact that such experiences can arise from deep unhappiness as well as a feeling of peace:

> One night of, I should think, neurotic misery I suddenly had an experience as if I was buoyed up by waves and waves of utterly sustaining power and love. The only words which came near to describing it were 'underneath are the everlasting arms' though this sounds like a picture, and my experience was not a picture but a feeling, and there were the arms.
>   …it came from outside unasked. No wishful thinking was involved, my unhappiness did not matter if the world was sustained by love in that way.[28]

The accounts collated by the Religious Experience Research Unit are autobiographic—people's recollections of times in their lives. Almost all are rare, often 'one-off' experiences, often of great peacefulness or joy—something of the feeling of deep joy that Lyra in *Northern Lights* gets when she rides Iorek Byrnison for the first time,[29] or that Lucy and Susan have when they play with Aslan in *The Lion, the Witch and the Wardrobe* after the scene at the Stone Table.[30]

Near-death experiences

Near-death experiences are profound personal experiences that typically happen when someone thinks they are about to die. People who have had near-death experiences subsequently often report such things as out-of-body experiences, various pleasant sensations and seeing a tunnel or light. Some claims associated with out-of-body experiences, such as being able to see what lay on shelves too high to be visible from the position of the body, do not seem to stand up to empirical tests.[31] Such accounts scarcely constitute conclusive evidence for the nature of the transcendent. Near-death experiences tell us nothing about what really happens when we die for the simple reason that those who relate such experiences have not died—they are near-death not death experiences. If it's not too dismissive, it's a bit like someone describing what it's like to be a Bedouin if the nearest they have got to a Bedouin is seeing them from afar whilst on a holiday in the Middle East. But a scientist might be interested in exploring *why* apparently large numbers of people have the sorts of experiences that Hardy found and that some people with near-death experiences describe.

Hardy's accounts, gathered for his Religious Experience Research Centre, are examples of retrospective data from people who have reported their experiences without having previously been studied.[32] The best studies of near-death experiences are prospective, which entail gathering a sample of people and then following them over time. Prospective studies have found that between 12 and 18 per cent of those who survive cardiac arrest subsequently report out-of-body experiences.[33] There is some evidence that young age is associated with the likelihood of having a near-death experience after a cardiac arrest and that women tend to have more intense near-death experiences than do men. Beyond those weak correlations, most studies find that nothing strongly predicts the likelihood of reporting a near-death experience—not ethnicity, marital status, occupational status, or religiosity.[34]

What can we learn from the observation that reports of near-death experiences are so widespread? Some of the reports are from people who are medically qualified, and therefore more likely to appreciate any neurophysiological basis of their experience.

Eben Alexander is a neurosurgeon. He spent 15 years on the faculty of Harvard Medical School. On Monday 10 November 2008 he woke an hour before the alarm clock was due to go off. He shifted slightly and a wave of pain shot down his spine. Within a few hours he was unconscious, convulsing violently and being rushed to hospital. It turned out that he had bacterial meningitis, with an infection of *Escherichia coli*— something that happens to fewer than one in ten million adults a year.

Alexander spent the next seven days in a coma. His conditioned worsened and family members who were not already present were summoned by the doctors as he was expected to die shortly. Remarkably, he opened his eyes—causing one of his visitors to shriek—and then started to reassure everyone, saying, 'All is well...Don't worry...all is well.'[35] Alexander went on to make a full recovery. He wrote about what he experienced during his week of unconsciousness and his subsequent interpretations of it:

> My meningoencephalitis had been so severe that my original memories from within coma did not include any recollections whatsoever from my life before coma, including language and any knowledge of humans or this universe. That 'scorched earth' intensity was the setting for a profound spiritual experience that took me beyond space and time to what seemed like the origin of all existence.

These memories began in a primitive, coarse, unresponsive realm (the 'Earthworm's Eye View' or EEV) from which I was rescued by a slowly spinning clear white light associated with a musical melody, that served as a portal up into rich and ultrareal realms. The Gateway Valley was filled with many earth-like and spiritual features: vibrant and dynamic plant life, with flowers and buds blossoming richly and no signs of death or decay, waterfalls into sparkling crystal pools, thousands of beings dancing below with great joy and festivity, all fueled by swooping golden orbs in the sky above, angelic choirs emanating chants and anthems that thundered through my awareness, and a lovely girl on a butterfly wing who proved months later to be central to my understanding of the reality of the experience (as reported in detail towards the end of my book Proof of Heaven). The chants and hymns thundering down from those angelic choirs provided yet another portal to higher realms, eventually ushering my awareness into the Core, an unending inky blackness filled to overflowing with the infinite healing power of the all-loving deity at the source, whom many might label as God (or Allah, Vishnu, Jehovah, Yahweh—the names get in the way, and the conflicting details of orthodox religions obscure the reality of such an infinitely loving and creative source).[36]

There has been some success in mapping near-death experiences to different parts of the brain with, for example, activity in the left hemisphere being associated with an altered sense of time and impressions of flying, and activity in the right hemisphere being associated with hearing voices, sounds, and music and seeing or communicating with spirits.[37]

There is little evidence in support of chemical-based theories though it is possible that the release of endorphins (hormones secreted within the nervous system) may play a role.[38] One theory that has found more support is that near-death experiences result from a lack of oxygen to the brain. Air pilots who have experienced unconsciousness during rapid acceleration subsequently sometimes describe experiences similar to those of near-death experience, such as seeing through a tunnel. Lack of oxygen may also trigger temporal lobe seizures which causes hallucinations, another common feature of near-death experiences.

Perhaps the best supported physiological explanation that we have at present for near-death experiences is the 'dying brain hypothesis'. This proposes that near-death experiences are hallucinations caused by activity in the brain as cells begin to die. However, this theory does not at present seem able to explain the full range of near-death experiences, such as why people often feel themselves to be out of their body.

Eben Alexander, whose remarkable recovery from bacterial meningitis we considered above, certain believes that his experience cannot be explained by natural forces and many—but not all—of those who provided accounts to Hardy's Religious Experience Research Unit believe similarly. Paul Badham, an emeritus professor of theology and religious studies, has spent much of his academic life writing about near-death experiences. He himself had one such experience when he was a teenager. While sensitive, as one might expect, to the danger of naively jumping from the occurrence of such experience to any sort of claim that they constitute a proof of God's existence, he notes that such experiences are common across a range of religions. He even goes as far as to claim that 'modern medicine's ability to resuscitate people who have "died" . . . is suggestive that historic beliefs in a real life beyond may have some evidential support.'[39]

From this brief survey of near-death experiences, we conclude that it would be a mistake to see them as a proof of something supernatural—beyond the sphere of science. Science may not at present have a complete and convincing natural explanation of such phenomena, though given their widespread occurrence this is likely to improve. However, this does not mean that we reject the possibility that such phenomena can help us discern the transcendent. *Brain-State Phenomena or Glimpses of Immortality?*, as one book title puts it, is a false dichotomy.[40] Just as there may well be natural explanations for the power of music or the awe that some natural phenomena cause us to feel, so near-death experiences too may both have natural explanations and be glimpses of the transcendent.

Accessing the transcendent

Natasha Ednam-Laperouse was allergic to sesame seeds. She was travelling to Nice with her father Nadim. While they were waiting at the airport he bought her a sandwich from a well-known chain. Nadim carefully checked the labelling for the ingredients, and deemed it safe for her. On the plane she became severely ill. Nadim injected her with adrenalin hoping that the reaction would subside, but things got worse and soon after landing paramedics applied CPR in vain. Shortly before Natasha died, Nadim had an extraordinary experience:

just as it couldn't get worse, these five angels just appeared, and thin yel
low light, strong, soft yellow light, rather like a candlelight but really
intense and in great detail, these five figures like thin people just appeared
with wings on their backs…They were about 20 cms tall, not chubby like
children in a Renaissance painting and with feathery wings like in the
Vatican, but actually like human beings, all looking at me, moving
around Natasha. I'd never ever seen anything like that in my life.[41]

Nadim and his wife Tanya have become well-known campaigners for
more accurate food labelling than that on the packaging of the
sandwich which proved fatal to Natasha. He says that at the time he was
an atheist, with no interest in religion. He has since started going to
church, and says that the material rewards of success no longer matter
as much to him as faith in God.

Such accounts do not gain credibility simply because they are associated with a death. No doubt they are amenable to neuropsychological
analysis. So is falling in love. And yet we would have an impoverished
view of human flourishing if we were to dismiss every avowal of romantic love as no more than a neuropsychological epiphenomenon. We
would also lose half of literature and much vocal music from opera to
pop songs. We can take a person like Nadim Ednam-Laperouse seriously
by doing our best to understand what his experience meant to him.

There are rare transcendent experiences in other walks of life. The
mathematician Sir Andrew Wiles was taught mathematics by Mary
Briggs, the mother of one of the authors. He wrote of her, 'Although it
is nearly forty-five years since I sat in her classroom I can still picture
the classes as if they were last year. She was an exemplary teacher whose
dedication to her students was quite exceptional. One remembers one's
teachers all one's life and how fortunate one is when the memories are
so good.'[42] Andrew Wiles is now best known for proving Fermat's last
theorem. In a press conference at the Heidelberg Laureate Forum in
September 2019 he described what it feels like to make a mathematical
discovery:

somehow you find this thing and suddenly you see the beauty of this landscape and you just feel it's been there all along. You don't feel it wasn't
there before you saw it, it's like your eyes are opened and you see it.[43]

When asked whether mathematics is discovered or invented, Wiles
replied that he didn't know a mathematician who doesn't think it is
discovered. The process of creativity in mathematics has much in

common with creativity in music.[44] If great mathematicians describe their work as discovery, what does that tell us about composing music and other creative activities such as poetry and painting?

Some creative people seem to have a sustained awareness of the transcendent. One such individual was the painter Stanley Spencer. The first volume of his published letters and diaries is appropriately called *Looking to Heaven*. In one respect, Spencer was the most rooted person one could imagine. His home village of Cookham on the River Thames in Berkshire meant so much to him that when he was an art student he was nicknamed 'Cookham' by his fellow students. Cookham features in many of his paintings. But in another respect Spencer was not of this world. Almost everything he does and sees, however mundane, is shot through with the divine, and that is the genius of his art.

Figure 4.5 shows Spencer's *Mending Cowls, Cookham*, which he painted in 1915. Cowls provide ventilation to oast houses where hops are dried. Spencer could see examples from the nursery window of Fernlea, the house in Cookham where he grew up. Of these cowls Spencer wrote:

> They seemed to be always looking at something and somewhere. When they veered round towards us they seemed to be looking at something above our own nursery window... With their white wooden heads, they served as reminders of religious presence.[45]

Figure 4.5   Stanley Spencer's 1915 painting *Mending Cowls, Cookham*.

To most of us, the cowls are impressive, even beautiful. But to Spencer they were more than that—'they served as reminders of religious presence'. A written example of Spencer's innate ability to integrate the transcendent and the immanent is provided by an extract from a long diary entry of his in 1917, when he was on active service in Salonica in the First World War, and thinking back to his beloved Cookham and the people he knows there:

> My friend has an Airedale terrier, a fine dog with a magnificent head neck and shoulders. He jumps leaps and bounds about in the dewy grass. I feel fresh awake and alive; that is the time for visitations. We swim and look at the bank over the rushes I swim right in the pathway of sunlight I go home to breakfast thinking as I go of the beautiful wholeness of the day. During the morning I am visited and walk about being in that visitations. How at this time everything seems more definite and to put on a new meaning and freshness you never before noticed. In the afternoon I set my work out and begin the picture. I leave off at dusk feeling delighted with the spiritual work I have done.[46]

This is painting as spiritual work—whatever the subject matter. It is in the same vein as William Blake, another great visionary and artist (and poet) whom we quoted at the start of the chapter:

> I ASSERT for myself that I do not behold the outward creation, and that to me it is hindrance and not action. 'What!' it will be questioned, 'when the sun rises, do you not see a round disc of fire somewhat like a guinea?' Oh! no, no! I see an innumerable company of the heavenly host crying 'Holy, holy, holy is the Lord God Almighty!' I question not my corporeal eye any more than I would question a window concerning a sight. I look through it, and not with it.[47]

It would be an impoverished person who denied appreciation of beauty on the grounds that it is all in the eye of the beholder. It would be comparably impoverishing to deny experience of the transcendent on the grounds that it is all in the spirit of the perceiver.

Enchantment

In 1920 the German sociologist Max Weber wrote a book whose title in English is *Sociology of Religion*. In it, he argued that Western society of the time was modern, bureaucratic, and secularized, in contrast to traditional society where 'the world remains a great enchanted garden'.[48] Weber therefore sees modernity as a time during which, while we gain much, we also lose much, including a sense of the sacred.

As one might expect, the pendulum swings back. It was another sociologist, Peter Berger, who in 1969 authored *A Rumour of Angels*, a book whose subtitle—*Modern Society and the Rediscovery of the Supernatural*—accurately summarizes its message. As Berger puts it in his concluding chapter:

> A rediscovery of the supernatural will be, above all, a regaining of openness in our perception of reality...In openness to the signals of transcendence the true proportions of our existence are rediscovered.[49]

The phrase 'signals of transcendence' is a fine one. It suggests another world keen to admit us, but a world that is generally veiled from sight. In his short story 'The Door in the Wall', H. G. Wells tells of a man, Lionel Wallace, who spends his life looking to recapture a time in his lonely childhood when he opened a door in a wall somewhere in West Kensington, London and found himself in another world:

> There was something in the very air of it that exhilarated, that gave one a sense of lightness and good happening and well-being; there was something in the sight of it that made all its colour clean and perfect and subtly luminous. In the instant of coming into it one was exquisitely glad—as only in rare moments, and when one is young and joyful one can be glad in this world. And everything was beautiful there....[50]

> Wallace mused before he went on telling me, 'You see,' he said, with the doubtful inflection of a man who pauses at incredible things, 'there were two great panthers there.... Yes, spotted panthers. And I was not afraid. There was a long wide path with marble-edged flower borders on either side, and these two huge velvety beasts were playing there with a ball. One looked up and came towards me, a little curious as it seemed. It came right up to me, rubbed its soft round ear very gently against the small hand I held out, and purred. It was, I tell you an enchanted garden. I know. And the size? Oh! it stretched far and wide, this way and that. I believe there were hills far away. Heaven knows where West Kensington had suddenly got to. And somehow it was just like coming home.[51]

A group of scholarly writers who used to meet for lunch on Tuesdays in the Eagle and Child pub in St Giles, Oxford, called themselves The Inklings. Their best-known members were C. S. Lewis and J. R. R. Tolkein. They were united in their conviction that the most basic carrier of truth is imagination. We shall say more about truth in Chapter 5. For now we note that a concept which they use to convey

what lies beyond the material world was 'magic'. This risks being mis
understood, but therein lies its memorability.

Ancient thinking about magic divided between animist beliefs in
which magic permeated nature and Platonist beliefs in which magic is
entirely outside nature. More recent ideas about magic demystify
nature, either with a separate but distinct heavenly deity or with no
deity and no magic at all. Some of the problems in thinking about God
arise from a stalemate between these older and newer modes of thought
about magic.[52] In our view none of them are correct.

A better country

Knowing God cannot be abstract. Descriptions of God are inseparable
from the narratives of those who claim to have encountered God.
That is why God has a name. In the Hebrew tradition God is often
referred to as the God of Abraham, Isaac, and Jacob.[53] When Moses was
responding to the somewhat challenging call to liberate the Hebrews
from slavery in Egypt, he insisted on having a name to use: 'Then
Moses said to God, "If I come to the people of Israel and say to them,
'The God of your fathers has sent me to you', and they ask me, 'What
is his name?' what shall I say to them?" '[54] The response was a name
which in Hebrew captures the fullest meaning of the verb 'to be': 'I AM
WHO I AM'.[55]

In the Christian tradition to which both the authors belong, there
are certain aspects of the encounter with God throughout history
which are almost universally recognized as core. They cannot be proved
to the satisfaction of an unrelenting sceptic, but the narrative is
supported by a substantial amount of documentary and other
evidence.[56] The story starts with the vision that humans should enjoy
an intimate relationship with God and the Creation. Because that did
not work out as intended, a plan was implemented whereby the
transcendent principle of the Universe became fused with the material
world as a human being and died in such a way that (whatever the
mechanism) the problem of human wrongdoing was somehow
addressed, and then appeared alive again never to die again.

In the New Testament book of Hebrews, the author of which is
unknown, there is in Chapter 11 a long litany of those who are seen as
having been faithful to God's word, beginning with Abel and continu-
ing to the birth of Christianity. In the middle of the list, we read:

These all died in faith, not having received what was promised, but having seen it and greeted it from afar, and having acknowledged that they were strangers and exiles on the earth. For people who speak thus make it clear that they are seeking a homeland. If they had been thinking of that land from which they had gone out, they would have had opportunity to return. But as it is, they desire a better country, that is, a heavenly one.[57]

This is a sense of the transcendent as pointing towards a world to come. We said earlier that attempts in earlier ages to prove God's existence nowadays no longer ring true to most people. But the sense that there is 'a better country', while completely unconvincing as a proof to the sceptic, is an encouragement to the one who already hopes for a reality beyond that which can be observed day in, day out and studied through the sciences.

In *The Penultimate Curiosity*, Andrew, with co-author Roger Wagner, argues that the desire by humans to make sense of the world, to ask ultimate questions about origins and meanings, gave rise in pre-history to a search for what might be termed 'penultimate questions', namely questions about the here and now, the purview of the sciences.[58] It can also work the other way around. Many scientists find that the more they study the natural world, the more they find it calling them to something beyond itself. The sciences can take us so far, but then they seem to point to what lies outside their field of operation.

The transcendent dimension of human flourishing

The three dimensions of flourishing which we have identified carry vast heterogeneity in their enjoyment and the choices that can be made. Throughout history there have been those with a surplus of material goods, and generally vastly larger numbers of people with a serious material deficit. For those with a surplus the challenge has been to make wise choices about their deployment; for those with a deficit the challenge has been to meet their own needs and those of their family and neighbours. For those in power, the challenge is to promote a combination of justice and generosity.

Leo Tolstoy reckoned that 'All happy families are alike; each unhappy family is unhappy in its own way.'[59] While that may be a succinct summary of the plot of *Anna Karenina*, there is a greater diversity in happy families than a literal reading of Tolstoy would allow. There are all sorts of ways in which humans can enjoy relational flourishing, just as there are severe cases of relational poverty. As with the material dimension of flourishing, there are varying degrees of control which the individual can exercise. For some, there may be genetic and/or mental impediments to forming relationships, or legacies of childhood abuse. For others the key to relational flourishing may lie in a decision to forgive. As Nelson Mandela so memorably expressed it, 'Resentment is like drinking poison and then hoping it will kill your enemies', or as Buddhaghosa, the fifth-century commentator on Buddhism, put it, 'Holding onto anger is like grasping a hot coal with the intent of harming another.'[60] Relational flourishing depends in a complex way on the lives we lead and the choices we make as to how we are going to relate to others.

The transcendent dimension is somewhat different to the material and relational dimensions. Some would argue that humans can flourish without it. At a global level the contribution of religion to human flourishing is contentious, with some arguing that religion is at the basis of most people's personal morality (and so responsible for a lot of good in the world) and others that religion is the cause of many of the world's great evils.[61] Insofar as religion describes a human endeavour one might expect both to be true. There are plenty of people who seem healthy and satisfied with their lives yet express no interest in or involvement with religion. There are some who have, at best, only an attenuated sense of the transcendent—for example, if they not only have no religious beliefs and practices but also, although they may 'quite like' the out-of-doors or 'quite enjoy' music, do not find anything in nature or the arts that is transformative, that takes them out of themselves.

For those who choose it, there is ample evidence that participation in religious events (such as attending worship) associates with improved physical and mental health and with feeling that one's life is more satisfying.[62] While religious participation may stimulate awareness of the transcendent, and even engagement with it, there is something about the transcendent that is too big to be limited to religious institutions and ceremonies. Sometimes the transcendent is about something more than the everyday breaking into our routine, whether this is a momentous event, like the birth of a child, or something less unusual but still important, like a beautiful sunset or a view of a wild animal that stops us in our tracks. But sometimes we need to make a deliberate decision to seek the transcendent. This may be through communal worship, often helped, no doubt, by the endorphins we discussed earlier, through prayer, through mediation, or through simply trying at all times to live conscious of eternity.

Notes

1 Miles, G. (2007) *Science and Religious Experience: Are They Similar Forms of Knowledge?*, Eastbourne: Sussex Academic Press.

2 Berry, R. J. (2014) *True Scientists, True Faith*, Oxford: Monarch Books.

3 Theophilus, N. A. (2016) A philosophical look at the egocentric interpretation of self-transcendence in man in the light of Nietzsche, *Journal of Philosophy, Culture and Religion* 23, 1–12.

4 Nietzsche F. (1883–5) *Also Sprach Zarathustra: Ein Buch für Alle und Keinen*, Chemnitz: Ernst Schmeitzner.

5 Addiction Centre (2020) Ecstasy addiction and abuse. Available at https://www.addictioncenter.com/drugs/ecstasy/.

6 Kalant, H. (2001) The pharmacology and toxicology of 'ecstasy' (MDMA) and related drugs, *Canadian Medical Association Journal* 165(7), 917–28.

7 Fans of the series *Star Trek* or its sequels and spin-offs.

8 Fandom, Memory Alpha: Nexus. Available at https://memory-alpha.fandom.com/wiki/Nexus.

9 Evans, J. (2017) *The Art of Losing Control: A Philosopher's Search for Ecstatic Experience*, Edinburgh: Canongate Books.

10 Dunbar, R. I. M. (2020) Religion, the social brain and the mystical stance, *Archive for the Psychology of Religion* 42(1), 46–62.

11 Oppy, G. (2019) Ontological arguments, *The Stanford Encyclopedia of Philosophy*, February 6. Available at https://plato.stanford.edu/entries/ontological-arguments/.

12 Irvine, A. D. (2020) Bertrand Russell, *The Stanford Encyclopedia of Philosophy*, May 27. Available at https://plato.stanford.edu/entries/russell/.

13 Russell, B. A. W. (1967) *The Autobiography of Bertrand Russell*, London: George Allen and Unwin, vol. 1, p. 133.

14 Plantinga, A. (2000) *Warranted Christian Belief*, Oxford: Oxford University Press.

15 Pippard, A. B. (1981) Accounts of the origin of matter, *The Times*, 21 December, 9.

16 Crick, F. (1994) *The Astonishing Hypothesis: The Scientific Search for the Soul*, London: Simon and Schuster, p. 3.

17 James, W. (1902) *The Varieties of Religious Experience: A Study in Human Nature*, New York: Longmans, Green & Co.

18 Goodman, R. (2017) William James, *The Stanford Encyclopedia of Philosophy*, October 20. Available at https://plato.stanford.edu/entries/james/.

19 James (1902) *The Varieties of Religious Experience*, p. 515.

20 Harrison, P. (2015) *The Territories of Science and Religion*, Chicago: University of Chicago Press, p. 1.

21 Harrison, P. and Roberts, J. H. (Eds) (2019) *Science without God?*, Oxford: Oxford University Press.

22 Morgan, E. (1982) *The Aquatic Ape*, New York: Stein & Day.

23 Hardy, A. (1960) Was man more aquatic in the past?, *New Scientist* 7(174), 642–5.

24 Alister Hardy Religious Experience Research Centre, https://www.uwtsd. ac.uk/library/alister-hardy-religious-experience-research-centre/.

25 Rankin, M. (2005) *An Introduction to Religious Experience*, Religious Experience Research Centre, Lampeter, p. 11. Available at https://repository.uwtsd.ac. uk/474/1/RERC3-002–1.pdf.

26 The Alister Hardy Trust, Personal Stories, 000322. Available at https://www. studyspiritualexperiences.org/personal-stories.html.

27 Ibid., 000071.

28 Ibid., 000356.

29 Pullman, P. (1995) *Northern Lights*, New York: Scholastic Point.

30 Lewis, C. S. (1950) *The Lion, the Witch and the Wardrobe*, London: Geoffrey Bles.

31 Marsh, M. (2010) *Out-of-Body and Near-Death Experiences: Brain-State Phenomena or Glimpses of Immortality?*, Oxford: Oxford University Press.

32 Sleutjes, A., Moreira-Almeida, A. and Greyson, D. (2014) Almost 40 years investigating near-death experiences: an overview of mainstream scientific journals, *Journal of Nervous and Mental Disease* 202(11), 833–6.

33 Van Lommel, P. (2014) Getting comfortable with near-death experiences: Dutch prospective research on near-death experiences during cardiac arrest, *Missouri Medicine* 111(2), 126–31.

34 Blanke, O., Faivre, N. and Dieguez, S. (2015) Leaving body and life behind: Out-of-body and near-death experience, in S. Laureys, O. Gosseries, and G. Tononi (Eds), *The Neurology of Consciousness*, 2nd edn, Cambridge MA: Academic Press, pp. 323–47.

35 Alexander, E. (2012) *Proof of Heaven: A Neurosurgeon's Journey into the Afterlife*, London: Piatkus, p. 113.

36 Alexander, E. (2012–19) My experience in coma. Available at http:// ebenalexander.com/about/my-experience-in-coma/.

37 Dagnell, N. and Drinkwater, K. (2018) Are near-death experiences hallucinations? Experts explain the science behind this puzzling phenomenon. *The Conversation*, 4 December. Available at https://theconversation.com/are-near-death-experiences-hallucinations-experts-explain-the-science-behind-this-puzzling-phenomenon-106286.

38 Wolfe, J. F. (2015) 10 Scientific explanations for near-death experiences, *Listverse*, 14 April. Available at https://listverse.com/2015/04/14/10-scientific-explanations-for-near-death-experiences/.

39 Badham, P. (2013) *Making Sense of Death and Immortality*, London: SPCK, p. 57. See also Badham, P. (1997) Religious and near-death experience in relation to belief in a future life, *Mortality* **2**(1), 7–21. Available at https://core.ac.uk/download/pdf/96773532.pdf.

40 Marsh, M. (2010) *Out-of-Body and Near-Death Experiences: Brain-State Phenomena or Glimpses of Immortality?*, Oxford: Oxford University Press.

41 Buchan, E. (2019) *A Bright Yellow Light*, BBC Radio 4. Available at https://www.bbc.co.uk/programmes/articles/1c575Zkjg7RDmy3Hgd01KrP/a-bright-yellow-light.

42 Wiles, A (2009) Email to Andrew Briggs to be read at the funeral of Mary Briggs.

43 Wiles, A. (2016) What does it feel like to do maths?, *+plus magazine*. Available at https://plus.maths.org/content/andrew-wiles-what-does-if-feel-do-maths.

44 Mcleish, T. C. B. (2019) *The Poetry and Music of Science: Comparing Creativity in Science and Art*, Oxford: Oxford University Press.

45 Tate Britain (2019) Sir Stanley Spencer, *Mending Cowls, Cookham 1915*. Available at https://www.tate.org.uk/art/artworks/spencer-mending-cowls-cookham-t00530.

46 Spencer, S. (2016) *Looking to Heaven* (J. Spender, Ed.), London: Unicorn Press, p. 265.

47 Blake, W. (1965) *The Poetry and Prose of William Blake*, ed. David V. Erdman, Garden City, NY: Doubleday, p. 555. Available at http://bq.blakearchive.org/14.1.strickland.

48 Weber, M. (1920/1971) *The Sociology of Religion*, London: Methuen, p. 270.

49 Berger, P. L. (1969/1970) *A Rumour of Angels: Modern Society and the Rediscovery of the Supernatural*, Harmondsworth, UK: Penguin, p. 119.

50 The two sets of four dots in this quotation are in the original.

51 Wells, H. G. (1927/1958) *Selected Short Stories*, Harmondsworth, UK: Penguin, pp. 109–10.

52 Tyson, P. G. (2019) *Seven Brief Lessons on Magic*, Eugene, OR: Cascade Books, p. 40.

53 Exodus 3: 6 and *passim*.

54 Exodus 3: 13.

55 Exodus 3: 14.

56 Wright, N. T. (2019) *History and Eschatology: Jesus and the Promise of Natural Theology*, London: SPCK.

57 Hebrews 11: 13–16a.

58 Wagner, R. and Briggs, A. (2016) *The Penultimate Curiosity: How Science Swims in the Slipstream of Ultimate Questions*, Oxford: Oxford University Press.

59 Tolstoy, L. (1878/2017) *Anna Karenina*, trans. Rosamund Bartlett, Oxford: Oxford University Press.

60 Paraphrased from the *Visuddhimagga*.

61 Dawkins, R. (2006) *The God Delusion*, New York: Bantam Book. Hitchens, C. (2007) *God is not Great: How Religion Poisons Everything*, New York: Twelve Books. Gill, R. (2018) *Killing in the Name of God: Addressing Religiously Inspired Violence*, London: Theos. Available at https://www.theosthinktank.co.uk/cmsfiles/Killing-in-the-Name-of-God.pdf.

62 VanderWeele, T. J. (2017) On the promotion of human flourishing, *Proceedings of the National Academy of Sciences* **114**(31), 8148–56.

PART II

PILLARS OF HUMAN FLOURISHING

Overview

Human flourishing cannot float in the abstract without any visible means of support. It requires moral and intellectual foundations. In a BBC programme entitled *Desert Island Discs*, every castaway is offered the Bible and the complete works of Shakespeare. The Bible contrasts two houses with and without strong foundations.[1] When the storm came, only the house built on secure rock remained standing. Shakespeare revised Hamlet's famous speech to include, in the second quarto edition, 'the slings and arrowes of outragious fortune', which were so severe as to tempt Hamlet to end his own life.[2] Without any visible means of support, human flourishing is just a matter of opinion, susceptible to the 'storm' or 'slings and arrowes' of fickle fluctuations in intellectual and moral standards, robust only insofar as a consensus can be found in any culture at any time. In Part II we seek to establish a firmer underpinning in terms of truth, purpose, and meaning.

A connecting theme running through Part II is that each of this triumvirate of *truth*, *purpose*, and *meaning* depends on human choices. That may surprise some readers, especially starting with *truth*. Writing in a post-truth culture at a time when fake news seems to be threatening our whole understanding of information in a networked age (as though there were ever a time before fake news), surely we should be asserting that truth is truth, whoever you are and whatever you think. You are entitled to your own opinions, but not to your own facts.[3] We wholeheartedly endorse the objective reality of truth, but that does not mean that it is independent of the observer.

In the sciences, knowledge is often expressed in terms of critical realism. We hope that our knowledge describes reality, but we accept that such a description is provisional, and liable to be updated as science advances. It is unhelpful to assert that Einstein showed that Newton's theories were untrue; it is more faithful to the way science works to say that Einstein's theory took Newton's work to deeper levels of understanding, and to higher levels of accuracy. No scientific theory is forever the truth, the whole truth, and nothing but the truth. But a good scientific theory may be sufficiently true to be useful. Whether it is useful depends on the question being asked, the problem to be solved, or the engineering design to be undertaken. These are all human choices.

In quantum science, the dependence of a measurement outcome on the choice of what to observe takes on a new level of mathematical formalism, with an associated set of philosophical conundrums.[4] This observer-dependence of truth is reflected in other disciplines too. In the nineteenth century the distinguished German historian Leopold von Ranke provoked a discussion about history as a description of what actually happened (*wie es eigentlich gewesen*).[5] Whatever the meaning of *eigentlich*, which can range from 'really' to 'essentially', it is not possible to write history simply as a series of facts. In German, *Geschicte* can be used for 'what happened', and *Historie* for its interpretation. But *Geschicte wie es eigentlich gewesen* is not achievable. Even if history could consist simply of a video recording, the film-maker would need to choose which direction to point the camera and whose voice to record. This is similar to the choices made by a cartographer; the only fully accurate map is a 3-D print of the complete terrain, including every blade of grass. Anything less reflects choices about what features to represent.[6] There can be bad histories and inaccurate maps, but there are no histories and no maps that are completely true. Critical realism separates the reality of something from a critical examination of how we know what we know. Our knowledge of a given reality depends on our engagement with it, and is provisional and subject to revision. A distinguished theologian and New Testament scholar has extended the concept of critical realism to the humanities including history.[7] The role of choice in truth thus spans our dimensions of flourishing from the material to the transcendent.[8]

Purpose can often be seen as an objective reality. The cosmologist Paul Davies was awarded the Templeton Prize in 1995. In his acceptance address he used the word purpose seven times, in some cases alongside meaning or design.[9] He concludes his book *The Mind of God* by reflecting that in humans a part of the Universe has become aware of the whole. He is convinced that the Universe has a purpose and that purpose includes us. He stops short, however, of saying what might lie behind that purpose; for him the Universe itself is the harmonious expression of a deep and purposeful meaning.

Within our human world, we are familiar with the concept of purpose in the context of a goal to be achieved. 'Why did you fix that hook to the wall?' 'Because I wanted somewhere to hang my coat.' The purpose in fixing the hook to the wall was to achieve the goal of having somewhere to hang the coat. Only the most reductionist determinist would deny the element of choice and decision-making in the whole affair. I decided that it was better to have somewhere to hang my coat than not to have somewhere to hang my coat. I decided on the place to fix the hook, and the choice of hook, and the method of fixing. There may have been all sorts of circumstantial background and constraints— my upbringing always to hang my coat on a peg; my repeated frustration at having no means to implement that; the layout of my hallway; the material of the walls; the selection available at my hardware store—but ultimately the purpose of fixing the hook was up to me. It was a choice that I made and for which I was responsible. The relationship between free will and responsibility is profound—maybe the former is a necessary but not sufficient condition for the latter—but it is hard to see how humans can flourish without being responsible.[10]

Thoughtful people often talk about finding a purpose for their lives. It is especially welcome when they do this before they are thirty years old. You can understand how to give content for such a quest. What are the person's abilities? In which academic subjects did they score highly? At what other activities did they excel? What obsessed them when no one was watching? What are their detailed memories of something which they found to be of enduring satisfaction, when they were 7 years old? 14 years old? 21 years old? 28 years old (if they have got there)? What are their burning ethical concerns? All these are questions which a good counsellor might ask. And yet, when all this information is to hand, it is the individual who will have to decide. There is an element of *finding* the purpose for our lives; there is a much stronger element of *choosing* the purpose for our lives, which may even involve *creating* it. And what is true for individuals is also true for communities and societies.

Meaning is closely related to purpose, but can somehow extend further. Is meaning an objective reality? If someone has said or written something that we do not understand, we can ask about their meaning, as a way of clarifying our understanding. This is a kind of Bayesian process, in which the additional explanation can serve to increase the confidence and accuracy of our belief about the meaning they intended to convey. An interpreter can enable us to understand the objective meaning of what would otherwise be foreign to us.

The meaning of something can be richer than that. Sometimes we speak of meaning in terms of consequences. A high school student gets good grades in their exams. That means, says their teacher, that you will be able to go to university. A patient receives the result of a scan. That means, says their oncologist, that there will be a need for a programme of treatment. A motorist approaches a red traffic light. That means, says the Highway Code, that you must stop and wait until the light changes to green. In all these cases the meaning of the information is independent of its reception by a human subject, although the subsequent unfolding of events may depend on how the recipient chooses to respond to the information.

In a limited sense of what are the implications, meaning can be impersonal. When the oil pressure drops in a car engine, it means that there is insufficient lubrication and the engine will be damaged if it continues to run. But when two people kiss, there is an intentional communication which cannot readily be put into words. It demands a response, which can range from rejection to embrace.

The poet Stephen Spender relates how an undergraduate once asked Eliot: 'Please, sir, what do you mean by the line: "Lady, three white leopards sat under a juniper tree"?' Eliot looked at him and said: 'I mean, "Lady, three white leopards sat under a juniper tree."'[11] Spender observes that this was less than helpful.

In its fullest sense, meaning requires a creator and a recipient. We see this in every great creation, whether literature, poetry, art, music, dance, or any other form of human expression. The dancer who was asked the meaning of her performance replied well, 'If I could have put it into words, I would not have danced it.' The engagement between creator and recipient may be dynamic, which is why many people prefer live performances to recordings, but it does not need to be. Few would dismiss a favourite painting as meaningless, but even fewer would dare to specify with unlimited precision what the meaning is. And while we may talk loosely about what a film or piece of music 'does for us', the meaning conveyed is inextricably bound up in how we choose to respond. Choice plays an integral role in meaning.

All this is true for the authors of this book. Both of us believe passionately in the objectivity of truth, while recognizing our responsibility for choosing what questions to ask in pursuit of that truth.

Each of us, in our own way, has found ourselves changing direction in our professional lives more than once, needing to work hard to choose and create the purpose for our careers. Both of us believe profoundly that life is meaningful, while recognizing that the meaning requires engaging with its source.

Andrew Briggs, when he was coming to the end of his undergraduate degree at Oxford, was advised by the careers service that he would be suited for a career either as a patent lawyer or as a military pilot. He pursued neither option. Instead he spent two years as a schoolteacher, followed by a doctorate in physics and a degree in theology. As one senior faculty who knew him well put it, his choice was then 'professor or bishop'? At the time there was insufficient evidence that either would be an eventual outcome, and it remains a source of amazement to him that he ended up as the former. Michael Reiss was once advised to consider a career in naval architecture—even though he gets seasick very easily and has very poor visual-spatial ability. He, too, is not a bishop, though he is ordained in the Church of England, and he also is a professor. Although the career of professor is hard to choose as a purpose at the macro scale, at the year-to-year and micro timescales there can be few careers outside the creative arts and parenthood where the individual has greater freedom and responsibility for choosing what purpose to pursue. And both of us, through our faith commitment, have chosen what we count on as truth and where we look for meaning in our lives.

Notes

1 Matthew 7: 24–7; Luke 6: 47–9.

2 Shakespeare, W. (1604) *Hamlet*, Act III, Scene 1.

3 'Everyone is entitled to his own opinion, but not to his own facts.' Attributed to Pat Moynihan, https://en.wikipedia.org/wiki/Daniel_Patrick_Moynihan (accessed 30 October 2019).

4 Briggs, A., Halvorson, H. and Steane, A. (2018) *It Keeps Me Seeking: The Invitation from Science, Philosophy, and Religion*, Oxford: Oxford University Press, Chapters 7, 8.

5 Von Ranke, L. (1973) *The Theory and Practice of History*, trans. W. A. Iggers and K. von Moltke, Indianapolis: Bobbs-Merrill; 'Man hat der Historie das Amt, die Vergangenheit zu richten, die Mitwelt zum Nutzen zukuenftiger Jahre zu belehren, beigemessen: so hoher Aemter unterwindet sich gegenwaertiger Versuch nicht: er will blos zeigen, wie es eigentlich gewesen.'

6 This discussion is explored further in Kei Miller's highly acclaimed, prized-winning poetry collection: Miller, K. (2014) *The Cartographer Tries to Map a Way to Zion*, Manchester: Carcanet.

7 Wright, N. T. (2019) *History and Eschatology: Jesus and the Promise of Natural Theology*, London: SPCK, pp. 95–105.

8 McGrath, A. (2001–3) *A Scientific Theology*, 3 vols, Edinburgh: T&T Clark.

9 Davies, P. (1995) Physics and the mind of God: The Templeton Prize Address, *First Things*. Available at https://www.firstthings.com/article/1995/08/003-physics-and-the-mind-of-god-the-templeton-prize-address-24.

10 Briggs, Halvorson, and Steane (2018), *It Keeps Me Seeking*, Chapter 5.

11 Spender, S. (1966) Remembering Eliot, *The Sewanee Review* **74**, 58–84.

5

Truth

Perceptions of reality

Two great physicists of the twentieth century, Niels Bohr and Albert Einstein, never reached agreement about what is true in quantum theory. Einstein believed passionately that a quantum state should be regarded as something real, and that you ought to be able to make a true statement about its value. Bohr reckoned that the truth about a quantum state only emerged after you had performed an experiment, which would necessarily depend on a choice about what to measure. Einstein found that profoundly unsatisfactory. He wanted to be able to affirm the reality of the quantum state regardless of whether it had been measured. Bohr confined himself to, 'What can one say?' Einstein wanted to ask, 'What is really true?' Truth has proved elusive in quantum theory, and not only in quantum theory.

In the film *The Truman Show*, the lead character, Truman Burbank, unwittingly lives his life inside a 24/7 reality television show that has been running for his whole life. Truman thinks he lives inside a seaside town, Seahaven, which is actually an enormous Hollywood set with thousands of cameras to record Truman's every moment—from his birth. Scene after scene manages to combine humour and poignancy. The key point is that Truman begins to suspect there is something odd about his world. Eventually, he sails away and, despite the efforts of the show's creator and executive producer, his boat eventually bumps into what till then has literally been the edge of his world. Truman then finds an exit door and escapes into the real world.

In an English court of law, witnesses are required to give a solemn undertaking to tell the truth, the whole truth, and nothing but the truth. For practical purposes that may be serviceable to underline the seriousness of what is said, and the consequences of perjury. But the deeper you dig, the more problematic it is conceptually. 'I saw the accused at six o'clock.' 'With what uncertainty?', any trained scientist would immediately ask. If the witness does not also add complete descriptions of the person's location, gait, clothing, and facial expression, it is not the whole truth. Any inaccuracy, however incidental, introduces something other than the truth. A pedantic witness might add to their oath, 'insofar as it is relevant to the case in hand, and insofar as it is sufficiently accurate and reliable to inform and not mislead the court for the purposes of the decision which the jury is required to reach'. Those are rather substantial qualifications.

The more passionately we care about truth, the more careful we need to be in understanding its nature. As humans, our perception of truth is inseparable from our experiences and our ability to reflect on them and articulate them. In doing that, we must be prepared to answer the challenge, 'What leads you to believe that that is so?' In order to affirm the objectivity of truth, we therefore need to examine different ways of knowing what is true.

The Matrix

The Truman Show has similarities with Thomas More's 1516 *Utopia*, where More describes an island called Utopia (Latin for 'nowhere'), but it is closer to *The Matrix* films from 1999 onwards. Again, their key idea is that there are people living their lives in what they think is a reality. In *The Truman Show* only one person is being deceived. In *The Matrix* it's everyone—the rise of artificial intelligence has produced machines that have imprisoned humanity in a virtual reality system, though just occasionally there is something that causes those in the system to wonder if there is something amiss. In the first film of *The Matrix* trilogy, Morpheus, the leader of the rebels, offers Neo the choice between a red pill and a blue pill. The red pill will release Neo from the control of what is effectively a dream world—but the real world is far harsher and less certain than the virtual reality. The blue pill will lead him back to the comforts and non-reality of the Matrix. Neo chooses the red pill.

Could we be living in the sort of world envisaged by *The Truman Show* or *The Matrix*? Four years after *The Matrix* first hit our cinemas, philosopher Nick Bostrom published his 'simulation argument'.[1] Bostrom imagines a future with enormous amounts of computing power. In such a situation people might run detailed simulations of their forebears or people like their forebears. Because the computers are so powerful, it is possible to imagine that the 'people' in the simulations might presume that their lives are the real ones. Bostrom goes so far as to conclude that this is probably the case   i.e. that most of us are living in simulations with only a minority (in our futures) living in reality.

The cosmologist Paul Davies has attempted to evaluate the likelihood that we are living in a simulated universe. A simulated universe would be less costly in energy than a real universe (although not free, because of the energy cost of information). Those who espouse a multiverse theory might reasonably argue that simulated universes should be more plentiful than real universes, and hence that we are more likely to be living in a simulated than in a real one. In that case, he asks, what guarantee is there that those in charge of it will continue to run it? They might get bored, or (a particular fear for those of us in science research) run out of funding! Or they might decide to vary the experiment and abruptly change the rules:

> While it may be true that our universe *is* a fake, it seems to me that drawing that conclusion would spell the end of scientific inquiry. In that respect it is akin to the argument that the universe was created five minutes ago with all the records and memories imprinted on it—that the present is real, but the past is fake. We could not disprove this claim, but accepting it gets us nowhere.[2]

While speculation that we are living in a cosmic simulation may provide an entertaining pastime for philosophers, it has limited operational value. And yet writers through the ages have warned against regarding current human experience as the only reality, or even the supreme reality. An anonymous first-century author described religious practice as containing but a shadow of the good things to come, not the true picture.[3] George Frederick Handel immortalized (if that is an appropriate verb in this context) in the *Messiah* what Paul wrote to the church in Corinth, 'The trumpet shall sound, and the dead shall be raised incorruptible, and we shall be changed. For this corruptible must put on incorruption and this mortal must put on immortality.'[4] So far from being pie in the sky when you die, this provides an incentive for practical action now. The scholar and former Bishop of Durham Tom Wright observes:

> Paul, we remind ourselves, has just written the longest and densest chapter in any of his letters, discussing the future resurrection of the body in great and complex detail. How might we expect him to finish such a chapter? By saying, 'Therefore, since you have such a great hope, sit back and relax, because you know God's got a great future in store for

you'? No. Instead, he says, 'Therefore my beloved ones be steadfast, immovable, always abounding in the work of the Lord, because you know that in the Lord your labour is not in vain.'[5]

In his inimitable creative style, C. S. Lewis imagines a bus ride to heaven in *The Great Divorce*.[6] To convey the sense that everything there is more real than anything the narrator has previously experienced he makes the light more intense, the trees more solid, and the grass stiffer, so much so that humans are ghosts by comparison. In this description Lewis was using a conceptual vocabulary with Greek origins.

Plato's Cave

The issues raised by *The Truman Show*, *The Matrix*, and Nick Bostrom's simulation argument have similarities with one of the foundational arguments in Western philosophy, Plato's Cave, which we briefly mentioned in Chapter 1. Plato's argument finds echoes in Zhuangzi, who lived about a hundred years after him in the Warring States Period in China:

> Once upon a time, I dreamt I was a butterfly, fluttering hither and thither, to all intents and purposes a butterfly. I was conscious only of my happiness as a butterfly, unaware that I was myself. Soon I awaked, and there I was, veritably myself again. Now I do not know whether I was then a man dreaming I was a butterfly, or whether I am now a butterfly, dreaming I am a man.[7]

The allegory of Plato's Cave is illustrated in Figure 5.1. Plato (through his mouthpiece, Socrates) envisages most people as being like prisoners who are in a cave, chained, and unable to turn their heads. Ahead of them is one of the walls of the cave; behind them, at the back of the cave, is a fire. Between the fire and the backs of the prisoners there is a walkway. On this walkway, puppeteers (unbeknown to the prisoners, as they are unable to look behind them) hold up puppets whose shadows are cast on the wall of the cave ahead of the prisoners. The prisoners view these shadows and presume that what they see is reality.

Plato then imagines that one prisoner is freed, turns around, and sees the fire. The light from the fire now makes it difficult for him to see the puppets that are casting the shadows, as he isn't used to bright lights. Chances are, he would therefore reject any explanation as to what causes his reality (to us, the shadows) on the wall in front of the

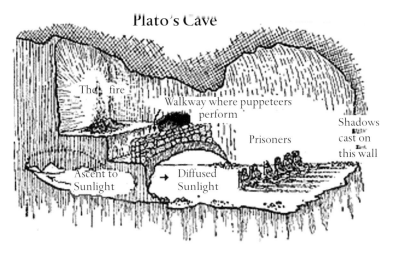

Plato's Cave

Figure 5.1 A contemporary visual representation of the allegory of Plato's Cave which Plato uses to argue that it is not straightforward to determine the nature of reality.

prisoners. Suppose further that someone takes him to the back of the cave and ascends with him to (what we know is) the outside world, where there is bright sunlight. It would take him a while to adjust to this bright light and then even longer to work out what is going on. Finally, Plato supposes that this freed prisoner returns to the cave. Would the other prisoners be convinced by his account? Or would it seem too fanciful? What would they choose to believe?

There are a number of ways of interpreting Plato's Cave. What any of us thinks is true may not turn out to be so. It can be difficult to discern the truth. There are different types of truth and different methods of inquiry are needed to establish the truth in different fields of knowledge.

Representations of truth

Every field of knowledge aspires to articulate the truth. Any such representation necessarily reflects the interests of the inquirer, and the purpose for which they are seeking the truth. This may be imposed on them by others, or it may be a choice which they freely make themselves.

In 1884 the Oxford Historical Society published a set of facsimiles of maps of Oxford spanning 150 years from 1578 to 1728. To anyone familiar

with Oxford the maps at first look all wrong, until they learn that until
1750 it was normal for maps to be drawn with south upwards. There is
an element of convention in the representation of truth. Despite the
differences in the dates of the maps, there is an encouraging consistency
between them. The castle, the city wall and its gates, the colleges, and
the churches are all in roughly the same places in the maps. But what
does it mean to ask whether maps are true?

Every map reflects the interests of the mapmaker and the purpose
for which it was drawn. A map for ramblers needs to show footpaths
and a certain amount of topographical information. A road map need
show neither of those, but it must show the different classes of roads
and the details of their intersections. A street map of a city for
prospective house-buyers needs to show the roads and their names but
not much more, whereas a map of the same city for utility companies
needs to show the location of all the underground pipes and cables. An
aviation map needs to show airfields, airways, and controlled airspace,
and many other features which might be of little interest to those on
the ground. For aviation purposes it needs to be completely reliable;
lives depend on it. Does that make it true?

The Canadian scholar and journalist John Stackhouse suggests that
truth is a quality, not a thing. Someone inspired by Plato's allegory of
the cave could not actually visit and witness an entity called truth
somewhere in the cosmos. Rather, truth is a quality of an interpretation,
so that a map is true to the extent that it accurately corresponds to the
reality that it represents:

> Truth is a quality of interpretations or representations by which we
> denote the extent to which they resemble reality—both in an absolute
> sense (according to their correspondence) and in a relative sense of their
> usefulness in that task (according to their pragmatic value). A map is
> more or less true in these two ways.[8]

Every map involves selection and approximation, and is therefore at
best only imperfectly true. A fully complete map would have to
reproduce every nanometre, indeed every molecule, of the territory
being mapped. Such a map might be perfect in its accuracy, but not
useful in its deployment, as Lewis Carroll once pointed out: 'We actually
made a map of the country, on the scale of a mile to the mile!...It has
never been spread out, yet...the farmers objected: they said it would
cover the whole country, and shut out the sunlight! So we now use the

country itself, as its own map, and I assure you it does nearly as well.'[9] A practical map needs to be sufficiently true to be useful for the purpose for which it is intended. More generally, knowledge needs to be sufficiently true for the purpose for which it is intended; we might even say for the purpose for which it is chosen. For many (but not all) purposes knowledge is often segmented into disciplines, each with its own criteria for truth.

Aspects of truth can be said to be scalable if they are found to apply in a wide range of circumstances. Good fiction is valued if it elucidates universal observations about aspects of human nature. Good poetry is valued if it resonates with shared human experiences. Good history is valued if lessons from the past seem to be applicable to understanding the present and deciding about the future. Good science is valued when a limited number of principles describe a wide range of phenomena. That is why the tools of mathematics are so powerful in science. In a lecture delivered in New York in 1959 the Hungarian-born Nobel laureate Eugene Wigner marvelled at 'the unreasonable effectiveness of mathematics in the natural sciences'. He used the word 'miracle' twelve times, and concluded:

> The miracle of the appropriateness of the language of mathematics for the formulation of the laws of physics is a wonderful gift which we neither understand nor deserve. We should be grateful for it and hope that it will remain valid in future research and that it will extend, for better or for worse, to our pleasure even though perhaps also to our bafflement, to wide branches of learning.[10]

Wigner was careful to write about the *effectiveness* and the *appropriateness* of mathematics. What about the *truth* of mathematics?

Mathematical truth

Most of us know that the sum of the internal angles of a flat triangle is 180°. A few of us may even remember one or more proofs of this, perhaps from our school mathematics lessons. More interesting for our purposes than these proofs is what mathematicians do *not* do to determine the sum of the internal angles of a flat triangle. One of the things they do not usually do is proceed as many decent empirical scientists would. That is to say, they do not head out into the field, gather representative samples of triangles, bring them back to the lab, culture them, as accurately as possible determine for each triangle the magnitude of its three internal angles and then sum them—arriving at some answer like a mean of 179.7° with a standard deviation of 0.8°.

Mathematicians establish truth in different ways from natural scientists, to whom we shall turn in the next section. This book is not a mathematical treatise so we will confine ourselves to one example— showing how mathematicians can show that there are an infinite number of prime numbers. First, we need to clarify the proposition that we wish to prove. For our purposes 'infinite' in the phrase 'infinite number of prime numbers' simply means that there is no upper end to the number of prime numbers. If you think there are only a million prime numbers, I can come up with more. If you think there are a billion prime numbers, I can come up with more. A prime number is a whole, positive number (i.e. a 'natural number' like 2 or 7) greater than 1 that is not the product of two smaller natural numbers. So, for example, 6 is not a prime number as it is the product of 2 and 3, whereas 5 is a prime number.

Euclid was an Ancient Greek mathematician who worked around 300 BCE. He produced in his book *The Elements* the earliest proof we know of that shows that that there are an infinite number of prime numbers. Euclid used a kind of argument called in Latin *reductio ad absurdum*—which we introduced in Chapter 4 as showing that a given starting point leads logically to a conclusion which is absurd. It runs like this. Suppose you think that you have already found the largest prime number there is. To make it specific, let's suppose that is the number 11. You then take all the prime numbers up to and including 11, and multiply them together: $2 \times 3 \times 5 \times 7 \times 11 = 2,310$. Now add 1 = 2,311. If you try to divide that number, 2,311, by what you thought were all the prime numbers, you will get a remainder of 1 in each case, so it is not divisible by any of those. As it happens, 2,311 is prime, so our original supposition that the largest prime number was 11 is false. If we had taken as our largest known prime the number 13, our procedure would have yielded $2 \times 3 \times 5 \times 7 \times 11 \times 13 + 1 = 30,031$. Now $30,031 = 591 \times 509$, so it is not itself prime, but it is the product of two prime numbers each of which is larger than our previously supposed largest prime number. Euclid generalized this to say that either the new number is prime or it is the product of other prime numbers not included in the earlier list. Either way the previous supposition that there was a largest prime number is proved to be false—*reductio ad absurdum*.

You may find this account that there are an infinite number of prime numbers convincing or you may feel that we have rather skated over certain matters when we wrote 'Euclid generalized this to say that either the new number is prime or it is the product of other prime numbers not included in the earlier list.' One of the great achievements of mathematicians has been to turn such statements, which can often lack precision, into rigorous mathematical arguments.

Problems with mathematical knowledge

For a long time, mathematicians assumed that by carefully tidying up their use of language and systematizing mathematical operations, they could put mathematics on an unshakeable foundation. To a large extent mathematicians have succeeded in this, but a number of problems arose along the way and the consensus nowadays is that these problems are insurmountable. The first of these problems is now known as Russell's paradox. Inevitably, for anyone who is not a strong mathematician it's challenging to follow the reasoning involved—though a heroic and readable attempt is made in the graphic novel *Logicomix*.[11] The philosopher and mathematician Bertrand Russell made a key discovery in 1901 concerning set theory. Using everyday language, a set is a collection of 'things' (real or imagined). Let us define a particular set, R, as the set of all sets that are not members of themselves. Now let us ask whether R is a set that is not a member of itself. Suppose the answer is 'yes'. Well, we defined R as 'the set of all sets that are not members of themselves', so this means that the answer 'yes' cannot be correct because we would have the logical impossibility that R is both a member of itself and not a member of itself. Unfortunately, we reach the same logical impossibility if we suppose that the answer to the question as to whether R is a set that is not a member of itself is 'no'. For many years in Wolfson College, Oxford, there was a grand piano in the dining hall which had a notice placed on top of it saying 'DO NOT PUT ANYTHING ON THIS PIANO'—a visual illustration of Russell's paradox.

We suspect that most readers will not find that their lives have suddenly lost all meaning on hearing of this paradox. However, mathematicians took it badly. Their presumption was that key areas of mathematics—and set theory is one such key area—could be firmly established, not shown to have logical contradictions at their core.

Things got worse for mathematicians as the twentieth century went on. The next key contribution (if that isn't putting it too positively) was made by Kurt Gödel in 1931, when he published his two incompleteness theorems at the age of 25. Gödel's incompleteness theorems are even harder for a non-mathematician to understand than is Russell's paradox, though a fine attempt is made by Douglas Hofstadter in his Pulitzer Prize-winning *Gödel, Escher, Bach: An Eternal Golden Braid*.[12] The essence of Gödel's work is that he managed to prove that if one can come up with a body of mathematics (technically, a computable axiomatic system) that can cope with the arithmetic of the natural numbers (1, 2, 3, 4, and so on—there is a separate discussion as to whether zero is a natural number)—which is a pretty modest achievement—then two rather surprising conclusions follow:

1. If the system is internally consistent, it cannot be complete.
2. The consistency of axioms cannot be proved within their own system.

Because this is mathematics, terms such as 'internally consistent' and 'complete' are used in very precise ways. As one might expect, there is an enormous literature on the implications of all this for mathematics. For our purposes, the important conclusion that follows from the work of Russell, Gödel, and others is that while mathematics remains robust and extremely useful, it is not as logically consistent and complete as many suppose.

Many people find the proofs of mathematics to be elegant and beautiful, even with the above caveats. However, there is much truth outside the scope of mathematics. Scientific truth adds to reason the test of experiment.

Scientific truth

Fundamentally, science is concerned with the natural world and with certain elements of the manufactured world—so the laws of gravity apply as much to artificial satellites as they do to planets and stars. Science is concerned with how things *are* rather than with how they *should* be. So there is a science of gunpowder and of in vitro fertilization without science telling us whether or when warfare and test-tube births are good or bad.

Attempting to define what science is and how it is undertaken is not straightforward. The sociologist Robert Merton characterized science as open-minded, universalist, disinterested, and communal.[13] For Merton, science is a group activity; even though certain scientists work on their own, all scientists contribute to a single body of knowledge accepted by the community of scientists. The nineteenth-century French physiologist Claude Bernard expressed it as 'L'art c'est moi, la science c'est nous.' Individual scientists are often passionate about their work and sometimes slow to accept that their cherished ideas are wrong. But science itself is not persuaded by such partiality. Time (almost) invariably shows which of two alternative scientific theories is nearer the truth, as we shall illustrate below in the case of the steady-state theory versus the Big Bang theory in cosmology. For this reason, if for no other, scientists are well advised to be open-minded, prepared to change their views in the light of new evidence or better explanatory theories.

As we intimated in Chapter 1, one of the key features of scientific theories, hypotheses, and conclusions is that they are falsifiable, a criterion emphasized by the philosopher Sir Karl Popper.[14] Unless you can imagine collecting data that would allow you to refute a theory, the theory isn't scientific. So the hypothesis 'All swans are white' is scientific because we can imagine finding a bird that is manifestly a swan (in terms of its appearance and behaviour) but is not white. Indeed, this is precisely what happened when early white explorers returned from Australia with tales of black swans (Figure 5.2). Popper has been described as one of the greatest philosophers of science of the twentieth century.[15] He developed his ideas in reaction to what he perceived as pseudoscience in the Viennese culture in which he grew up. He had in mind the political theories of Marx and Engels and the psychoanalytic theories of Freud and Adler. What appeared to be a strength of those theories, namely their ability to accommodate and explain every possible form of human and social behaviour, was in fact, he argued, a critical weakness, because it followed that they could not robustly predict anything. They were therefore untestable. By contrast, Einstein's theory was risky, because it made prediction which at the time seemed utterly counterintuitive, and which, for example, could have been falsified by Eddington's observations of the apparent gravitational deflection of light from a star by the Sun during the total eclipse of 1919.[16]

Popper's ideas have been challenged in their detail, though the basic concepts remain highly influential. As so often, it is one thing to write a scheme on a whiteboard; it is an altogether different matter to implement it within all the vagaries of experimental uncertainty. It is rare for a single observation to falsify a theory, because there are always possibilities that the observation is mistaken or that the assumed

Figure 5.2 Karl Popper pointed out that no number of white swans is sufficient to prove the assertion that all swans are white. A single non-white swan, though, is enough to disprove the assertion.

background knowledge may be faulty or incomplete. Most of the time Popperian thinking is far removed from how scientists actually go about their tasks. Few scientists enter their laboratory each morning wondering, 'What theory can I falsify today?' They are much more likely to be hoping that they can get the experiment to work!

Problems with knowledge in the natural sciences

Popper's ideas can give rise to a view of science in which knowledge steadily accumulates over time as new theories are proposed and new data collected to discriminate between conflicting theories. This rather optimistic view of the inexorable growth of scientific knowledge has been undermined by more recent historians, sociologists, and philosophers of science who have pointed out that while science may search for universal truths about the material world, a whole raft of factors mean that this goal is far from being attained.

At the simplest level, most science requires funding—often lots of it. Louis Pasteur, the French biologist and chemist who invented the process of pasteurization, said that fortune favours only the prepared mind,[17] a maxim which might now be restated as fortune favours only the funded mind! Funding decisions are seldom made purely in pursuit of universal scientific knowledge. Even when 'blue skies' funding is approved, there are often deep-seated interests at work—think of national interests with regards to scientific exploration, whether to new lands in the eighteenth century, or to the poles or space in the twentieth century.

The American Moon landing in 1969 was a direct response to the competition between the USA and the USSR. 'Do we have a chance of beating the Soviets by putting a laboratory in space, or by a trip around the moon, or by a rocket to go to the moon and back with a man? Is there any other space program that promises dramatic results in which we could win?' Kennedy had asked his staff in an internal memo.[18] Science funding in advanced countries still has a place for unfettered discovery, but there is an inexorable trend towards government priorities and initiatives. Governments generally have a complex mixture of motives, from promoting human flourishing as they see it to getting re-elected.[19]

Alongside funding constraints, scientists are humans. They often stick to favoured theories longer than the evidence warrants,[20] they sometimes massage their findings, and their youthful passion for objective truth can with age metamorphose into a quest for tenure, money, status, and plaudits. Given time, such human shortcomings should wash out—one of the great strengths of science is its capacity to autocorrect. Nevertheless, there is an increasing realization that scientific plagiarism and even fraud are more frequent than had generally been acknowledged.[21]

It is very difficult to establish the extent of fraud in science. One highly cited study found that 2 per cent of scientists admitted to have fabricated, falsified or modified data or results at least once and concluded that this figure was probably a conservative one.[22] It can be difficult to establish whether fraud has taken place. Cyril Burt (1883–1971) was an educational psychologist who played an important role in developing the '11-plus' examination in schools in England to determine whether students would be educated from the age of 11 in more (grammar schools) or less (secondary modern) academically demanding schools, a decision with life-long consequences for each child. Although it is generally thought that Burt systematically engaged in scientific fraud, falsely claiming to have collected data in his studies on the heritability of intelligence,[23] there have been revisionist accounts, defending his integrity.[24]

Scientific malpractice can have severe consequences. From 1996 to 2013, Yoshihiro Sato, a Japanese bone-health researcher, plagiarized work, fabricated data, and forged authorships. As a result, more than 60 studies in the academic literature have been retracted.[25] Sato's fraud is thought to have been one of the biggest in scientific history—it is certainly one of the largest that has come to light. Much of Sato's purported research was on how to reduce the likelihood of bone fractures. Because of his fraud, meta-analyses that included his 'data' came to the wrong conclusion, medical organizations produced erroneous guidelines, and in follow-ups on studies that they did not know were faked, other researchers carried out new trials that enrolled thousands of patients.[26] Even top journals frequently carry retractions of papers, which simultaneously provides evidence of the problem and of the extent to which the checks and balances of the international science community make it difficult for fraud to go undetected for long.

In principle, one of the tests of the truth of scientific work is whether it can be repeated. In practice, life is often too short, and scientists have too many other priorities, to test everything in that way. A scholarly process which is almost universally applied to publications in reputable scientific journals (indeed, academic journals in general) is peer review. Peer review is a human, and therefore fallible, endeavour, but it is almost indispensable in science publishing. This then provides a sociological criterion for truth: has the paper been peer-reviewed? The reason it works as well as it does in science is that behind peer review lies the final court of appeal: does it work? That may be what a scientist ultimately means by 'is it true?' In the natural sciences established results are normally fully repeatable. Where they are not, then the ideas do not generally survive for long. They are denied the coveted epithet of true. What is not true cannot provide a basis for human flourishing.

Truth in the social sciences

The social sciences include such disciplines as psychology, sociology, economics, anthropology, and education. Some social sciences have sub-disciplines that, to all intents and purposes, belong to the natural sciences. Experimental psychology, for example, involves studying the behaviour of humans and other animals using the same sort of experimental design and the same sort of statistical analyses that are used in the natural sciences. But some social sciences ask questions that are very different to those asked in the natural sciences.

Consider the apparently innocent question 'What is the optimal number of school students in a class?' It turns out that once one begins to attempt to answer this question, value judgements are inexorably entwined. For a start, is a successful outcome to be equated with maximum knowledge acquired or with something quite different, such as the likelihood of students being sufficiently engaged to choose to continue with the subject once it is no longer compulsory? There is no objective way of deciding between these alternatives, because the choice depends on values.

Then, even assuming we have come to a decision on this issue, should we consider not just the students in the class but other students? After all, any slight benefit to the students in one class in a school in having their numbers reduced means that at least one other class in the school will need to have more students. If we are talking about decreasing average class size across a large geographical area (such as a state or nation), then correspondingly more teachers will be required overall. If there is a shortage of good teachers—and there nearly always is—these additional teachers may not on average be as good as those already in post.

Then there is the observation that teachers not only prefer smaller classes but are more likely to leave the profession if they have larger classes.[27] We could go on. Questions in the social sciences are not only often more complex (in the sense that there are more variables to consider and these cannot always be disentangled) and more subject to value judgements than questions in the natural sciences, they are often interested in knowledge that is more affected by local circumstances. The boiling point of water may be virtually the same in Lagos, London, and Tokyo, but the best teaching methods are not.

Even within a single class, not all students learn best in the same way. Consider how best to teach children to read (Figure 5.3). In many countries what have been referred to as 'reading wars' have been raging for many years. On the one side is 'whole language instruction'. This means using actual books. The books might be written by their authors

Figure 5.3 Children reading at a primary school in rural Laos in 2013. There isn't a single best way of learning to read that works for all learners. The best way for any of us to read depends on who we are, what we are trying to read and the characteristics of our teacher, if we have one.

simply to engage their readers or they might exist within a reading scheme that carefully increases vocabulary, sentence length, plot complexity, etc. over a series of books. On the other side of the proponents of whole language instruction are the proponents of phonics.

Phonics is a way of teaching reading (and writing) by developing an awareness in learners of the various phonemes that make up an oral language. A phoneme is a particular unit of sound and a grapheme is the shortest meaningful unit in a writing system. So, for example, in English the phoneme 'b' (as defined by the International Phonetic Alphabet) corresponds to the graphemes 'b' and 'bb'—as, for example in the words 'bed' and 'bubble'. The phoneme 'eɪ' (an example of a falling diphthong because it starts with a vowel of higher prominence and ends with a short vowel) is informally known as 'long a' and is the sound we hear in such words as 'bay', 'maid', 'weigh', 'straight', 'pay', 'foyer', 'eight', 'gauge', 'mate', 'break', and 'they'. It therefore corresponds, in these words, to the graphemes 'a', 'ai', 'eigh', 'aigh', 'ay', 'er', 'ei', 'au', 'a_e', 'ea', and 'ey'.

Proponents of phonics hold that the best way to enable young children to read is to teach them the phonemes in a language—there are usually held to be forty-four phonemes in English—and how these phonemes correspond to graphemes. There is considerable evidence to show that in English-speaking countries, teaching by phonics on average enables more children to reach a certain standard of reading by a certain age than does whole language instruction on its own. But the key point is 'on average'. The two authors of this book both learnt to read by whole language instruction, and don't seem to have suffered too much as a consequence (but let the reader judge!). The social sciences often have to grapple with issues of subjectivity and relationships between people in ways that, at the very least, muddy attempts to discern universal truths.

Problems with knowledge in the social sciences

The social sciences face a greater range of problems with knowledge than do mathematics or the natural sciences. We have already indicated some of these in our discussion of optimal class size and the best way to teach reading where issues of values, subjectivity, and relationships between people give rise to considerable problems for anyone seeking after truth. A rather more down-to-earth problem in the social sciences, though it affects the natural sciences too, goes under the name of the 'reproducibility crisis' (or 'replication crisis').

The reproducibility crisis was first highlighted in psychology about a decade ago.[28] The key issue is easy to state. It has increasingly been found that many findings in such disciplines as psychology and medicine are difficult or impossible to reproduce. This is despite the fact that being able to reproduce a finding is held, as we have seen, to be a key feature of experimental science.[29] In 2016, the journal *Nature* asked 1,576 scientists across a range of disciplines for their thoughts on reproducibility. Over 70 per cent said they'd tried and failed to reproduce the result of someone else's experiments.[30] Indeed, just over half of them replied that they had tried and failed to reproduce their own!

It should not be assumed that this issue is one that only affects flaky work done by underfunded social psychologists. The crisis may have started with psychology but recognition of it has spread to other disciplines. The *Nature* survey above found that chemists were more likely than other scientists to report a failure to reproduce results—both others' and their own—while one very highly cited study that focused on fifty-three 'landmark' studies in preclinical cancer research reported that in only six of them (11 per cent) could the findings be confirmed.[31] What is going on?

Setting aside outright fraud, which we discussed above and is thankfully generally thought to be rare, there is a whole host of reasons why experimental findings may not be reproducible. The *Nature* survey lists the follow in descending order of importance:

- Selective reporting
- Pressure to publish
- Insufficient oversight/mentoring
- Low statistical power or poor analysis
- Not replicated enough in original lab
- Poor experimental design
- Methods, code unavailable
- Insufficient peer review
- Raw data not available from original lab.

In recent years, a whole raft of procedures have been tightened up and new ones introduced in various disciplines in attempts to address the reproducibility crisis. It's probably too early to say with confidence how successful these are proving but the whole episode should serve as a reminder that in both the social and the natural sciences there is a provisionality about conclusions, in ways that vary both between disciplines and within them. How we can steer a course between blind trust in science on the one hand and throwing our hands up in despair and resorting to horoscopes and Tarot cards on the other is something to which we return below in the section on postmodernism.

By scientific truth, we mean the well-established findings of the natural sciences such as physics, chemistry, and biology. Certain things clearly fall within science—optics, the arrangement of atoms into molecules, and vertebrate anatomy, to give three examples. However, what about the origin of the Universe, the behaviour of people in society, decisions about whether we should build nuclear power plants or go for wind power, the appreciation of music, and the nature of love? In ways that differ case by case these can be reduced to scientific issues, but is that the whole story? Some people would argue 'yes' and in Chapter 4 we explained, unapprovingly, that the term *scientism* describes the view that science can provide sufficient explanations for just about everything (except mathematics).[32] Indeed, at one point in the first half of the twentieth century, it was even seriously held by certain philosophers and others (such as the logical positivists who constituted the Vienna Circle in the 1920s and 1930s) that questions that could not be answered by either logical proof or empirical observation—questions, for instance, to do with ethics and aesthetics—were meaningless.

However, most people hold that there are other forms of knowledge alongside mathematical and scientific knowledge. This way of thinking means that the origin of the Universe is also a philosophical or even a religious question—or simply unknowable. In addition to the natural sciences, understanding the behaviour of people in society requires knowledge of the social sciences (e.g. psychology and sociology). Whether we should go for nuclear or wind power involves not only science and engineering and an understanding of economics, risk, and politics, but also ethical choices about future generations and the planet itself. The appreciation of music and the nature of love have something to do with our perceptual apparatuses and our evolutionary history, but the person who seeks to reduce them only to science is—well—missing something.

Moral truth

Are there moral truths or only moral opinions? Both possibilities raise problems. If there are only moral opinions, who are you to complain if I pull the wings off flies—or worse? If there are moral truths, where do they come from? A religious believer might say 'from God' but many people aren't religious believers. Anyway, there is the issue first recorded by Plato in the *Euthyphro*, where Socrates asks whether the good is loved by the gods because it is good or it is good because it is loved by the gods.[33] Can the good simply be what God defines it to be? Even a religious believer might consider that the good should have a degree of independence of God, just as something could be beautiful in itself rather than because a deity deems it to be. At the start of the Judaeo-Christian scriptures, at the end of each of the six days of Creation, God is recorded as seeing 'that it was good'—so that God recognizes goodness rather than defining it as good, although in this case the epithet may refer to fitness for purpose rather than moral goodness.[34]

Ethics is the branch of philosophy concerned with how we should decide what is morally wrong and what is morally right. Ethics is both a practice and a branch of knowledge. Ethical thinking is not wholly distinct from thinking in other disciplines but it cannot simply be reduced to them. Ethical conclusions cannot be unambiguously proved in the way that mathematical theorems can. However, almost everyone agrees that this does not mean that all ethical conclusions are equally valid. Some ethical conclusions are more likely to be valid than others.[35] How are we to discern ethical truth?

We can be most confident about the validity and worth of an ethical conclusion if three criteria are met.[36] First, if the arguments that lead to the particular conclusion are convincingly supported by reason. Secondly, if the arguments are conducted within a well-established ethical framework. Thirdly, if a reasonable degree of consensus exists about the validity of the conclusions, arising from a process of genuine debate.

It might be thought that reason alone is sufficient to justify an ethical conclusion in attempts to arrive at moral truths. However, there is no single universally accepted framework within which ethical questions can be decided by reason.[37] This is not to say that reason is unnecessary but to acknowledge that reason alone is insufficient. For instance, reason cannot decide between an ethical system which looks only at the consequences of actions and one which considers whether certain actions are right or wrong in themselves, whatever their consequences.

The criterion of a well-established ethical framework is susceptible to ethical drift, because the acceptability of the resulting ethics may become a criterion for the validity of the ethical framework within which they arose. If you think that Nazi ethics were not acceptable, then you may assert that this shows the Nazi ethical framework was invalid. The so-called 'No true Scotsman' argument runs thus: 'No true Scotsman puts sugar on his porridge.' 'My Scottish uncle puts sugar on his porridge.' 'That shows that he is not a true Scotsman.' Narrative can play a foundational role in an ethical framework. The traditional Hebrew ethic to respect resident aliens is based on the narrative of the Hebrew experience of themselves being aliens in Egypt.[38] Narrative can provide a strong basis for ethics, albeit seldom a universally accepted basis.

The insufficiency of reason and the absence of a widely accepted single ethical framework means that we are forced to consider the approach of consensus.[39] But consensus does not solve everything. After all, what does one do when consensus cannot be arrived at? Nor

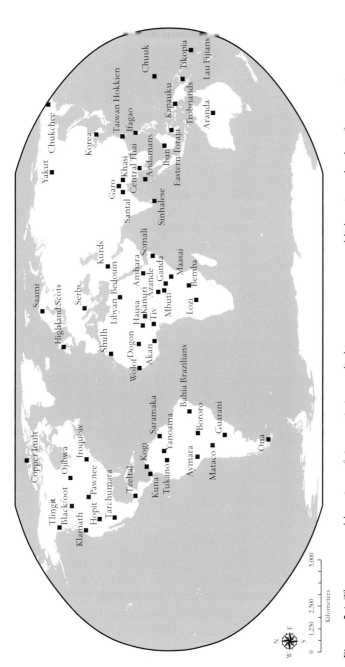

Figure 5.4 The names and locations of sixty societies studied in an attempt to establish universal rules of cooperation among people.

can one be certain that consensus always arrives at the right answer—in many societies a consensus once existed that beating was good for children, or that it was acceptable to discriminate against Jews, or that Black lives didn't matter as much as White lives.

Nevertheless, there are good reasons both in principle and in practice in searching for consensus. Such a consensus should be based on reason and genuine debate and take into account long-established practices of ethical reasoning. At the same time, it should be open to criticism, refutation, and the possibility of change. Finally, consensus should not be equated with majority voting. Consideration needs to be given to the interests of minorities, particularly if they are especially affected by the outcomes, and to those—such as young children, the mentally infirm, and non-humans—unable to participate directly in the decision-making process. It also needs to be borne in mind that while a consensus may eventually emerge, there is a time when what may be more important is simply to engage in valid debate in which the participants respect one another, so far as is possible, and seek for truth through dialogue.[40]

For all that there are widespread disagreements about what is morally right and what is morally wrong, there is much agreement too. One recent large-scale study began with the premise that morality consists of a collection of biological and cultural solutions to the problems of cooperation recurrent in human social life—the 'morality-as-cooperation' theory.[41] It then looked at the distribution of cooperative behaviours among the sixty societies shown in Figure 5.4. It found that seven behaviours—helping kin, helping your group, reciprocating, being brave, deferring to superiors, dividing disputed resources, and respecting prior possession—are widely valued, despite the considerable differences between these societies.

Religious truth

Mathematics, the natural sciences, the social sciences, and moral philosophy are all disciplines of knowledge. Religion is rather different—it consists of beliefs, practices, and experiences, with theology as its intellectual support. It is notoriously difficult to define in a way that includes what most people regard as all religions (Buddhism doesn't have gods, for instance) but excludes what most people do not regard as religions yet seem to get included by definitions that include all widely recognized religions (Marxism, for example). The following, derived from the work of Ninian Smart[42] and John Hinnells[43] are generally characteristic of most religions:[44]

- Religions have a *practical and ritual dimension* that encompasses such elements as worship, preaching, prayer, yoga, meditation, and other approaches to stilling the self.
- The *experiential and emotional dimension* of religions has at one pole the rare visions given to some of the crucial figures in a religion's history, such as that of Arjuna in the *Bhagavad Gita* and the revelation to Moses at the burning bush in Exodus. At the other pole are the experiences and emotions of many religious adherents, whether a once-in-a-lifetime apprehension of the transcendent we discussed in Chapter 4 or a more frequent feeling of the presence of God either in corporate worship or in the stillness of one's heart when solitary.
- All religions hand down, whether orally or in writing, vital stories that comprise the *narrative or mythic dimension*, for example the story of the Exodus in the Judaeo-Christian scriptures. For some religious adherents such stories are believed literally; for others they are understood symbolically.
- The *doctrinal and philosophical dimension* arises, in part, from the narrative/ mythic dimension as theologians within a religion struggle to integrate these stories into a more general view of the world. Thus, the early Christian church came to its understanding of the doctrine of the Trinity by combining the central truth of the Jewish religion—that there is but one God—with its understanding of the life and teaching of Jesus Christ and the working of the Holy Spirit.
- If doctrine attempts to define the beliefs of a community of believers, the *ethical and legal dimension* regulates how believers act. So Sunni Islam has its Five Pillars—*Shahada* (testimony of faith), *Salat* (prayer), *Zakat* (alms-giving), *Sawm* (fasting), and *Hajj* (pilgrimage to Mecca)—while Judaism has the Ten Commandments and other regulations in the Torah, and Buddhism has its Five Precepts.
- The *social and institutional dimension* of a religion relates to its corporate manifestation: for example, the Sangha (the order of monks and nuns founded by the Buddha to carry on the teaching of the Dharma) in Buddhism; the umma' (the whole Muslim community) in Islam; and the Church (the communion of believers comprising the body of Christ) in Christianity.

- Finally, there is the *material dimension* to each religion, namely the fruits of religious belief as shown by places of worship (e.g. synagogues, temples, and churches), religious artefacts (e.g. Eastern Orthodox icons and Hindu statues) and sites of special meaning (e.g. the river Ganges, Mount Fuji, and Uluru (Ayers Rock)).

Faced with this catalogue of religious dimensions, it can be quite difficult to get to grips with the notion of religious truth.

Attempts have been made to prove the truth of religion mathematically. A quick Internet search for 'mathematical truths in scripture' soon leads to a plethora of websites that claim mathematically to prove the existence of God. One titled 'The Mathematical Proof for Christianity Is Irrefutable' maintains that there are 300 fulfilled prophecies about the person of Christ—e.g. 'The Messiah will be born in Bethlehem. (Micah 5:2; Matthew 2:1; Luke 2:4-6)' and 'The Messiah will be betrayed by a friend. (Psalm 41:9; Luke 22:47,48)', and that the odds of all these prophecies being fulfilled are one in one hundred quadrillion, or 1 in 100,000,000,000,000,000.[45] Our view is that whatever one's view on biblical prophecy, such calculations are invalid—you simply can't work out such probabilities.

A different approach is to hold that the scriptures (particularly the Qur'an) contain mathematical truths about prime numbers or the value of $\pi = 3.14159\ldots$ (the ratio of the length of the circumference of a circle to the length of its diameter). Again, that is not how the two of us understand the way that scripture works. Related to this are the widespread beliefs that the Judaeo-Christian bible enables one to date the Earth (young Earth creationism) or that the Qur'an contains scientific truths about a whole range of topics that were unknown when it was written—such as the speed of light, the behaviour of black holes, and early embryonic development.

Many people, whether religious or not, when talking about religious truth may be thinking about the lessons for people as to how they should live their lives. These are lessons people learn, if they are religious, from their scriptures, from the recorded behaviours and teachings of the founders of their religion, and from the ongoing teachings and practices of their religious leaders and fellow believers.

Some, perhaps most, religions advocate that believers treat one another, or sometimes the whole of humanity, as they would members of their family. In the New Testament, Jesus is recorded as saying, 'Whoever does the will of God is my brother, and sister, and mother'.[46] The letters in the New Testament are full of references to fellow Christians as brothers and sisters and even as members of a single body: 'For as in one body we have many members, and all the members do not have the same function, so we, though many, are one body in Christ, and individually members one of another.'[47] In the early Church, the virtue of compassion and seeing all as needing God's grace equally so that, 'There is neither Jew nor Greek, there is neither slave nor free, there is neither male nor female; for you are all one in Christ Jesus' were taken seriously.[48] So much so that some historians believe that the theological virtue of love has to this day fundamentally shaped attitudes and social practices in the West.[49]

The Christian tradition, with which we the authors are most familiar, has from the beginning cared passionately about the truth. We shall return to this in connection with science before the end of this chapter. It has been forcefully expressed by the poet John Betjeman in his reflection on the Christmas story.

> And is it true? And is it true,
> > This most tremendous tale of all,
> Seen in a stained-glass window's hue,
> > A Baby in an ox's stall?
> The Maker of the stars and sea
> Become a Child on earth for me?[50]

There are plenty of other questions which can be asked of religious beliefs and practices. Some of them, such as the beneficial effects of prayer or forgiveness, are even amenable to empirical measurement of their contribution to human flourishing.[51] But underpinning all of these seems to be an inescapable question of the role of truth.

Seeking neither to tell the truth nor to lie

We are nearing the end of our survey of how truth is understood and attempts made in various disciplines to gain it. Before we consider our final discipline—that of postmodernism—we can note an essay published in 1986 with the unusual title, for a philosopher, of 'On bullshit'. In 2005 Harry Frankfurt's essay was published as a short book and proved a surprise success—helped no doubt by its title.

Frankfurt's central point is that there is a form of communication—bullshit—where the intention is to persuade without regard for truth. Those who try to tell the truth and those who intentionally tell untruths (i.e. lie) have in common that they have due regard for how things actually are. However, as Frankfurt puts it:

> The bullshitter...does not reject the authority of the truth, as the liar does, and oppose himself to it. He pays no attention to it at all. By virtue of this, bullshit is a greater enemy of the truth than lies are.[52]

The success of Frankfurt's book was due to far more than its title. In many ways, Frankfurt was ahead of his time. Some 35 years later, we are familiar with the idea that we live in a post-truth era. In 2016, 'post-truth' was named 'Word of the Year' by the Oxford Dictionary, which defines it as 'relating to or denoting circumstances in which objective facts are less influential in shaping public opinion than appeals to emotion and personal belief'.[53]

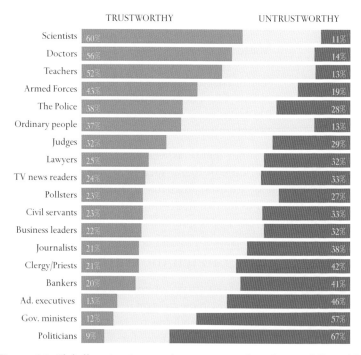

Figure 5.5 Globally, scientists are the most trusted profession, followed by doctors. Sample of 17,793 adults aged 16–64, interviewed October 2018.

Perhaps we can take some solace from the fact that when there are large-scale medical emergencies, like that arising from the spread of COVID-19, most people have a high regard for truth. We want to know how we can catch diseases, how to make it less likely that we will catch them, how serious they are, how they can be treated, and so on. It was the report of scientists of the Imperial College COVID-19 Response Team that changed UK government policy for coronavirus from mitigation to suppression.[54] The basis of the modelling has since been questioned, but that reinforces the extent to which those concerned care about truth. Post-truth politics seem like a luxury when something is a matter of life and death. Globally, scientists are the most trusted profession, followed by doctors (Figure 5.5).

Postmodernism

In 1990, a group of conservative philosophers attempted, unsuccessfully, to prevent one of the founding figures of postmodernism, Jacques Derrida, from being awarded an honorary doctorate at the University of Cambridge. They sent a letter to *The Times* in which they argued that his assertions were 'either false or trivial' and that his 'originality does not lend credence to the idea that he is a suitable candidate for an honorary degree.'[55]

Along with relativism—the idea that there are no absolute truths but that each of us can decide for ourselves—postmodernism is widely held to be an attack on truth. However, it can be seen more positively, and very differently from relativism. Postmodernism is an attempt to use critical and rhetorical practices to destabilize concepts such as certainty,[56] but it can also be seen as a rigorous attempt to discern rather than attack truth. Postmodernism is deeply suspicious of what are called metanarratives or grand narratives—overarching attempts to order knowledge and determine what is valid and what is not. Parenthetically, it is this rejection of a single way of determining what is best that has led to postmodern architecture in which colourful, playful buildings reject established canons of beauty (Figure 5.6).

The proponents of postmodernism see it not only as radical but as democratic. They are suspicious of Enlightenment rationality. We can illustrate a postmodern approach in the social sciences by looking at the work of Judith Butler, particularly in her celebrated book *Gender Trouble*.[57] Her fundamental argument is that gender is not an established reality. So she rejects the metanarrative that our gender naturally

Figure 5.6 Converted from a hospital, the Judge Business School (designed by John Outram) is one of the few postmodern buildings at the University of Cambridge. Causing a furore when built in 1995, it gained listed building status in 2018.

follows from our belonging either to the male or female sex. Rather, each of us—though Butler's focus is on women—learns throughout our lives to perform gender. In this, she builds on the earlier work of feminists such as Simone de Beauvoir. Butler argues that gender is performative and it is the performance that constitutes, rather than expresses, an illusion of stable gender identity. The far greater gender fluidity that we witness today among young people in the West follows from this. The rejection of the binary of 'male' and 'female' clothes has been extended far beyond how we dress.

Expectations of agreement about truth

Why do we sometimes fail to agree, even when presented with the same evidence? A possible explanation may lie in Bayesian *priors*. Bayesian probability is to be distinguished from frequentist probability, which is about the proportion of times you would expect a particular outcome if you repeat an experiment within given constraints. Bayesian probability is the probability on given evidence that a particular belief is true. We have to be careful here, because Bayes' theorem is a mathematical formula, and therefore calls for numbers to be put in. For some of the most important beliefs, such quantification is scarcely possible.[58]

Bayesian reasoning updates the prior belief (strictly, the probability that the prior belief is true) in the light of fresh evidence, through Bayes' theorem, which is rigorously derived by applying what is called 'the chain rule' to conditional probabilities. Crucially, the resulting belief depends on both the new evidence and the prior. Thus, if two people start with different priors, then logically they will end with different resulting beliefs.

In some circumstances, repeated application of Bayesian reasoning will eventually lead to a resulting belief that is almost independent of the prior. In that case, with sufficient evidence two people may converge on almost the same belief (strictly, the probability that a particular belief is true) even if they start with different priors. This is generally the case in science (we shall give an example below from cosmology). An analogy is two people trying to find the highest point of ground. If one person starts in France and the other in Switzerland, they may both converge on the top of Mont Blanc as the highest summit. But convergence is not inevitable. If one person starts in France and the other in Nepal, then they may disagree: one advocating Mont Blanc and the other Everest. Whether they can resolve their disagreement may depend on whether they have an agreed method of measuring absolute altitudes. The problem of local minima is well known in a number of fields including finding the lowest energy structure in materials modelling, and also in Bayesian optimization in machine learning. The machine may think (to anthropomorphize) that it has achieved the optimum, but only because it has not looked far enough. In Bayesian terms, it is possible for two people each to settle on different beliefs, each of which seems to one of them to have the highest probability of being true.

The key step is a shift from asking 'Is this belief true?' to asking 'What is the probability that this belief is true?' If two thoughtful people presented with the same evidence disagree about the probability that a particular belief is true, it is not necessarily because they are being irrational. It may be because they start with different prior beliefs in a context where the additional evidence does not lead to a convergence of revised beliefs.

The feasibility of agreement about truth varies according to discipline. In some subjects, vigorous discussion is generally conducted in the expectation that it will eventually lead to agreement, because there is some reality which the antagonistic proponents aspire to describe accurately (we might say sufficiently accurately to be useful), and for which there are agreed criteria of truth.

A major disagreement in cosmology in the twentieth century was about the origin of the Universe. There were powerful minds on both sides of the debate. Fred Hoyle, stalwart champion of the steady-state theory, took the stand for an infinite universe with no beginning and no end, in which matter is continuously created in the space between the galaxies. George Gamow, a principal architect of the Big Bang theory, made the case for a universe that began billions of years ago as an explosion from an extremely dense and infinitesimally small seed of energy. The steady-state theory had much to commend it, in terms of physical and mathematical elegance. It was a very attractive theory. Unfortunately for its advocates it proved to be wrong. Every cosmologist, astronomer, and physicist we know now accepts that the Universe had an origin in what has come to be known as the Big Bang, with an agreed date of approximately 13.8 billion years ago. The disagreement has to all intents and purposes been completely resolved.

People often talk of the Scientific Method, as if there were one such thing. It is often articulated as formulating a hypothesis on the basis of observations, and then carefully designing repeatable experiments, the outcome of which depends decisively on whether the hypothesis is true or false. Although this is a good starting point for scientific investigation, and can be recognized in much of the best science that is done, it is not nearly as universal as is sometimes supposed. Competing hypotheses about the origin of the Universe could not be tested by repeatable experiments, let alone control experiments with a single independent variable. But the resolution of the controversy did depend on science's secret weapon, if not of experiment, at least of observation. The clincher was an observation which no one was looking for, namely the cosmic background radiation for which Arno Penzias and Robert Wilson were awarded the 1978 Nobel Prize in Physics.

Contrast this with the level of (dis)agreement which might be expected in the arts or humanities. Take an example from opera. One person might assert that Verdi's *Rigoletto* is all about a father with a chip on his shoulder and a mixed-up daughter who love each other very much but who utterly fail to understand one another. Rigoletto cares a lot about Gilda, but cannot appreciate the emotional strength of her attraction to the count. Gilda is grateful to her father, but cannot recognize the wisdom of his judgement of character. The result is tragic for both of them. Another person might have a completely different interpretation of the opera, and see it as all about discrimination by the rich and powerful against a disadvantaged and disabled member of society. Such a disagreement will not be resolved by further data, since both people have the same libretto and score in front of them. The same could apply to the interpretation of almost any piece of literature, and to much in other humanities subjects too.

This difference in the expectation of agreement is not because science has a monopoly on the search for truth. Every intellectual discipline cares about truth. John Carey, emeritus Merton Professor of English Literature at Oxford, articulates the coincidence of purpose between such disparate disciplines as science and theology:

> science cannot be regarded as inherently anti-religious. On the contrary, its aims seem identical with those of theology, in that they both seek to discover the truth. Science seeks the truth about the physical universe; theology, about God. But these are not essentially distinct objectives, for theologians (or at any rate Christian theologians) believe God created the universe, so may be contacted through it.[59]

We are thus rescued (if we need to be rescued) from a kind of scientism that asserts that the Scientific Method (there is no such thing) or even scientific methods (slightly better) are the only means of discovering truth. Rather, we can conclude that different methods of inquiry are appropriate to different kinds of questions, and they may be capable of arriving at different levels of confidence that a particular belief is true, with different (and sometimes decisive) degrees of dependence on the prior belief with which a person embarks on the inquiry.

What is truth?

That question has been posed before.[60] The answer then is valid now. Perception of truth is inseparable from the attitude of the person who is seeking, or not seeking, truth. Even before the era of manipulating digital images and voices, a video recording of an incident told the viewer only about the scenes in the direction in which the camera was pointing, and the speech which was picked up by the microphone, and then only for the duration of time for which the machine was active. Excluding collateral data which might have been gathered, the truth of the record extended only as far as the purpose for which it was made.

Postmodernism makes us focus on issues of trust. Sometimes we can establish truths for ourselves but often we need to rely on others. Who are the experts in whom we can trust? This is not a straightforward matter. As the philosopher Julian Baginni puts it:

> We *need* to defer to experts but not everyone who claims to be an expert is one. If we decide which experts to defer to on the basis of expert opinion, we paradoxically have to choose which experts to trust in order to decide which experts to trust.[61]

Whom to trust is one of the most important kinds of choice that underpin human flourishing. Scientific inquiry and knowledge can and should inform decisions to trust, but if trust never goes beyond provable evidence then the resulting life will be severely impoverished. Hungarian-born Michael Polanyi was one of the great intellectuals of the twentieth century. He made lasting contributions to physical chemistry, economics, and philosophy. He was better qualified than most to appraise the variety of components of personal knowledge and the limits of objectivism:

> We owe our mental existence predominantly to works of art, morality, religious worship, scientific theory and other articulate systems which we accept as our dwelling place and as the soil of a mental development. Objectivism has totally falsified our conception of truth, by exalting what we can know and prove, while covering up with ambiguous utterances all that we know and *cannot* prove, even though the latter knowledge

underlies, and must ultimately set its seal to, all that we *can* prove. In trying to restrict our minds to the few things that are demonstrable, and therefore explicitly dubitable, it has overlooked the a-critical choices which determine the whole being of our minds and has rendered us incapable of acknowledging these vital choices.[62]

The extent to which human flourishing can be based on truth depends on the truth that humans choose to select. Those who study Shakespeare gain insights about human behaviour that might otherwise elude them. Those who study Maxwell's equations gain knowledge about electromagnetism that might otherwise baffle them. If that is the case about seemingly objective knowledge, how much more does it apply in the realm of relationships! One can learn a certain amount about someone from their biography or CV, but to uncover truth about another person at a deeper level requires engaging with them. In the process, as two volitional agents interact, not only the perception but also the substance of the truth about each of them may change.

How human flourishing builds on choices about truth depends in turn on choices about purpose and meaning, the subjects of our next two chapters.

Notes

1 Bostrom, N. (2003) Are you living in a computer simulation?, *Philosophical Quarterly* **53**(11), 243–55.
2 Davies, P. D. W. (2006) *The Goldilocks Enigma: Why is the Universe Just Right for Life?*, London: Allen Lane, p. 214.
3 Hebrews 10: 1.
4 1 Corinthians 15: 52–3.
5 Wright, N. T. (2007) *Surprised by Hope*, London: SPCK, pp. 204–5.
6 Lewis, C. S. (1946) *The Great Divorce*, London: Geoffrey Bles, p. 27 ff.
7 Wu, K.-M. (1990) *The Butterfly as Companion: Meditations on the First Three Chapters of the Chuang-Tzu*, New York: State University of New York Press.
8 Stackhouse, J. G. (2020) *Can I Believe? An Invitation to the Hesitant*, Oxford: Oxford University Press, p. 10 (slightly abridged).
9 Caroll, L. (1893/1982) *Sylvie and Bruno Concluded*, in *Lewis Carroll—The Complete Illustrated Works*, New York: Gramercy Books, p. 727.
10 Wigner, E (1960) The unreasonable effectiveness of mathematics in the natural sciences, *Communications on Pure and Applied Mathematics* **13**, 1–14.
11 Doxiadis, A. and Papadimitriou, C. H. (2009) *Logicomix: An Epic Search for Truth*, London: Bloomsbury.

12 Hofstadter, D. (1979) *Gödel, Escher, Bach: An Eternal Golden Braid*, New York: Basic Books.

13 Merton, R. K. (1973) *The Sociology of Science: Theoretical and Empirical Investigations*, Chicago: University of Chicago Press.

14 Popper, K. R. (1934/1972) *The Logic of Scientific Discovery*, London: Hutchinson.

15 Thornton, S. (2018) Karl Popper, *The Stanford Encyclopedia of Philosophy*, 7 August. Available at https://plato.stanford.edu/archives/win2019/entries/popper/.

16 A more accurate description would be in terms of the distortion of space-time by the mass of the Sun, with the propagation of light remaining straight.

17 Pasteur, L. (1854) *Dans les champs de l'observation le hasard ne favorise que les esprits préparés.* Lecture, University of Lille.

18 Kennedy, J. F. (1961) Memorandum for Vice President, 20 April, Presidential Files, John F. Kennedy Presidential Library, Boston, MA.

19 Collier, P. & Kay, J. (2020) *Greed Is Dead: Politics after Individualism*, London: Allen Lane.

20 Kuhn, T. S. (1970) *The Structure of Scientific Revolutions*, 2nd edn, Chicago: University of Chicago Press.

21 Reydon, T. A. C. (2020) What attitude should scientists have? Good academic practice as a precondition for the production of knowledge, in K. McCain and K. Kampourakis (Eds), *What Is Scientific Knowledge? An Introduction to Contemporary Epistemology of Science*, London: Routledge, pp. 18–32.

22 Fanelli, D. (2009) How many scientists fabricate and falsify research? A systematic review and meta-analysis of survey data, *PLoS ONE* 4(5): e5738.

23 Tucker, W. H. (1997) Re-reconsidering Burt: Beyond a reasonable doubt, *Journal of the History of the Behavioral Sciences* 33(2), 145–62.

24 Fletcher, R. (1991) *Science, Ideology, and the Media: The Cyril Burt Scandal*, Piscataway, NJ: Transaction Publishers. Tredoux, G. (2015) Defrauding Cyril Burt: A reanalysis of the social mobility data, *Intelligence* 49, 32–43.

25 Else, H. (2019) What universities can learn from one of science's biggest frauds, *Nature* 570, 287–8.

26 Kupferschmidt, K. (2018) Tide of lies: Researcher at the center of an epic fraud remains an enigma to those who exposed him, *Science* 361(6403), 636–41.

27 Isenberg, E. P. (2010) The effect of class size on teacher attrition: Evidence from class size reduction policies in New York State, US Census Bureau Center for Economic Studies Paper CES-WP-10–05.

28 Pashler, H. and Wagenmakers, E.-J. (2012) Editors' introduction to the special section on replicability in psychological science: A crisis of confidence?, *Perspectives on Psychological Science* 7(6), 528–30.

29 Kourany, J. A. (2020) What grounds do we have for the validity of scientific findings? The new worries about science, in K. McCain and K. Kampourakis

(Eds), *What Is Scientific Knowledge? An Introduction to Contemporary Epistemology of Science*, London: Routledge, pp. 212–25.

30 Baker, M. (2016) Is there a reproducibility crisis?, *Nature* **533**, 452–4.

31 Begley, C. and Ellis, L. (2012) Raise standards for preclinical cancer research, *Nature* **483**, 531–3.

32 Reiss, M. J. (2015) The nature of science, in R. Toplis (Ed.), *Learning to Teach Science in the Secondary School: A Companion to School Experience*, 4th edn, London: Routledge, pp. 66–76.

33 The original has 'τὸ ὅσιον' which can be translated by a number of words or phrases including 'the pious' and 'that which is sanctioned'. Its standard formulation ('the good') dates from Leibniz in the early eighteenth century.

34 Cottingham, J. (2009) *Why Believe?*, London: Continuum.

35 Reiss, M. J. (2019) Science, religion and ethics: The Boyle Lecture 2019, *Zygon* **54**(3), 793–807.

36 Reiss, M. (1999) Bioethics, *Journal of Commercial Biotechnology* **5**, 287–93.

37 O'Neill, O. (1996) *Towards Justice and Virtue: A Constructive Account of Practical Reasoning*, Cambridge, UK: Cambridge University Press. Parfit, D. (2011) *On What Matters*, vol. 1, Oxford: Oxford University Press.

38 Leviticus 19: 33–4.

39 Moreno, J. D. (1995) *Deciding Together: Bioethics and Moral Consensus*, Oxford: Oxford University Press.

40 Habermas, J. (1983) *Moralbewusstsein und Kommunikatives Handeln*, Frankfurt am Main: Suhrkamp Verlag. Martin, P. A. (1999) Bioethics and the whole: Pluralism, consensus, and the transmutation of bioethical methods into gold, *Journal of Law, Medicine & Ethics* **27**, 316–27.

41 Curry, S., Mullins, D. A. and Whitehouse, H. (2019) Is It good to cooperate? Testing the theory of morality-as-cooperation in 60 societies, *Current Anthropology* **60**(1), 47–69.

42 Smart, N. (1989) *The World's Religions: Old Traditions and Modern Transformations*, Cambridge, UK: Cambridge University Press.

43 Hinnells, J. R. (1991) *A Handbook of Living Religions*, London: Penguin Books.

44 Reiss, M. J. (2008) Should science educators deal with the science/religion issue?, *Studies in Science Education* **44**, 157–86.

45 Delzell, D. (2013) The mathematical proof for Christianity is irrefutable, *The Christian Post Opinion*, 28 May. Available at https://www.christianpost.com/news/the-mathematical-proof-for-christianity-is-irrefutable.html.

46 Mark 3: 35.

47 Romans 12: 4–5.

48 Galatians 3: 28.

49 Holland, T. (2019) *Dominion: The Making of the Western Mind*, London: Little, Brown.

50 Betjeman., J. (1958) 'Christmas', in *Collected Poems*, London: John Murray.

51 Worthington, E. L., Kurusu, T. A., McCollough, M. E. and Sandage, S. J. (1996) Empirical research on religion and psychotherapeutic processes and outcomes: A 10-year review and research prospectus, *Psychological Bulletin* **119**, 448–87.

52 Frankfurt, H. G. (2005) *On Bullshit*, Princeton, NJ: Princeton University Press, p. 61.

53 Oxford Languages (2016) Word of the Year 2016, Oxford University Press. Available at https://languages.oup.com/word-of-the-year/2016/.

54 Ferguson, N. M. et al. (2020) Impact of non-pharmaceutical interventions (NPIs) to reduce COVID19 mortality and healthcare demand. Available at https://www.imperial.ac.uk/media/imperial-college/medicine/sph/ide/gida-fellowships/Imperial-College-COVID19-NPI-modelling-16-03-2020.pdf.

55 Zabala, S. (2014) Ten years without Derrida, Aljazeera Media Network, 4 April. Available at https://www.aljazeera.com/indepth/opinion/2014/03/ten-years-without-derrida-20143291559170321.html.

56 Aylesworth, G. (2015) Postmodernism, *The Stanford Encyclopedia of Philosophy*, 5 February. Available at https://plato.stanford.edu/archives/spr2015/entries/postmodernism/.

57 Butler, J. (1990) *Gender Trouble: Feminism and the Subversion of Identity*, London: Routledge.

58 Briggs, G. A. D., Halvorson, H. and Steane, A. M. (2016) *It Keeps Me Seeking: The Invitation from Science, Philosophy, and Religion*, Oxford: Oxford University Press, pp. 153–8.

59 Carey, J. (1995) *The Faber Book of Science*, London: Faber, pp. xxii–xxiii.

60 John 18: 38.

61 Baggini, J. (2017) *A Short History of Truth: Consolations for a Post-Truth World*, London: Quercus, p. 29.

62 Polanyi, M. (1958/2015) *Personal Knowledge: Towards a Post-Critical Philosophy*, Chicago: University of Chicago Press, p. 286.

6

Purpose

Silent Running is a 1972 science fiction film that belongs to the post-apocalypse genre. Set in the future, the Earth is dying. As many plant species as possible are being kept in vast greenhouses set in space and tended by a crew that includes Freeman Lowell, the resident botanist. However, the crew receives an order to jettison the greenhouses and return the space ships that carry them so that they can be used for commercial purposes. Lowell rebels and enlists the support of his service robots (Figure 6.1). After a series of mishaps, Lowell is able to save one of the greenhouses, sacrificing himself in the process. The end of the film shows a greenhouse filled with flourishing plants and one of the service robots, Dewey, tenderly caring for them. Both Lowell and the robot Dewey have a clear purpose, though they differ in their agency, in their capacity to make and enact choices. The human capacity to make choices will be a theme to which we return in this chapter and the next.

Aristotle provides a route into clarifying what we mean by purpose by looking at the four types of causes that there can be. He recognizes four types of answer that can be given in response to a 'why?' question—such as 'Why is a bronze statue as it is?'[1] First (though there is no particular significance in the order), there is the *material cause*—that out of which the statue is made, namely 'bronze'. Second, there is the *formal cause*. For Aristotle, form is the account of what something is to be, so the form of a horse statue is 'a horse'. Third, there is the *efficient cause*—the primary source of what we are talking about. In the case of a statue, this is the artisan who literally pours the bronze and/or the sculptor who designed it. Finally, we have the cause over which more ink has perhaps been spilled than all the others combined, the *final cause*—the end, that for the sake of which that we are talking about has been made or done. In the case of a statue, this might be to

Figure 6.1   Freeman Lowell, the resident botanist, and two of the service robots in the film *Silent Running*.

commemorate a great warrior, to satisfy the creative urges of an artist or to sell for a profit.

Aristotle's final cause is concerned with purpose. We look at the Barbara Hepworth sculpture in front of the United Nations Secretariat Building in New York (Figure 6.2) and in answer to the question 'What is it for?', reply that it's a memorial to the former UN Secretary General Dag Hammarskjöld who was tragically killed in an air crash.

An institution such as marriage has a purpose and so do works of art in a sense—but what about the Universe? Does it have a purpose?

Figure 6.2 Barbara Hepworth's *Single Form*, her largest work (6.4 m in height), photographed at its unveiling ceremony at the United Nations Secretariat, New York, 11 June 1964. It was commissioned as a memorial to the UN Secretary General Dag Hammarskjöld after his death in an air crash in 1961. The material cause is bronze, the formal cause is an abstract sculpture, the efficient cause is Barbara Hepworth, and the final cause is a memorial to Dag Hammarskjöld.

The direction and purpose of the Universe

Whether or not the Universe has a purpose, it certainly has a history and it has a direction. Setting aside questions as to whether there is a multiverse, with our Universe being just one amongst many, and whether other universes have preceded ours or may follow its eventual death, the Universe is probably about 13.8 billion years old and probably started with what is commonly referred to as the Big Bang.[2]

The direction of the Universe can be understood in several senses. The British astrophysicist Arthur Eddington coined the phrase 'time's arrow' in his 1927 Gifford Lectures:

> Let us draw an arrow arbitrarily. If as we follow the arrow we find more and more of the random element in the state of the world, then the arrow is pointing towards the future; if the random element decreases the arrow points towards the past. That is the only distinction known to physics. This follows at once if our fundamental contention is admitted that the introduction of randomness is the only thing which cannot be undone.
>
> I shall use the phrase 'time's arrow' to express this one-way property of time which has no analogue in space.[3]

What is sometimes referred to as the thermodynamic arrow of time follows from the Second Law of Thermodynamics. Formally, but in non-mathematical language, this states that the total entropy of an isolated system can never decrease over time. 'Entropy' is the technical term for 'disorder' and one doesn't have to be a physicist to realize that things over time get disordered unless one puts effort into tidying them up. The reason why getting a teenager to tidy their bedroom doesn't contradict the Second Law of Thermodynamics is because the amount of additional order achieved by the tidying is more than cancelled out by the disorder created elsewhere. This is not a special property of teenagers. It is always the case that increasing order in one place is accompanied by at least the same of disorder being created elsewhere. This is the reason why the Second Law specifies 'an isolated system'. Our teenager's bedroom can (approximately) be considered as an isolated system but what may not be apparent to the naked eye is that the modest increase in order as a bed is made and a few clothes put away is more than outweighed by the disorder that results from respiration in the cells of the teenager's body—respiration being needed to keep teenagers alive and enable them to move their muscles—as relatively ordered molecules like glucose are broken down to simpler (more disordered) carbon dioxide and water.[4]

So, the Universe has a history (its past) and a direction (it can't go backwards); does it have a purpose? Biologists, certainly those who work on evolution or behaviour, tend to be perfectly comfortable with the idea that there is purpose in the lives of organisms—and we will turn to that in our next section—but physicists are more sceptical about the idea that the Universe has a purpose. The idea seems to be trespassing on the territory of religion. By and large scientists, whether they have a religious faith or not, tend to be pretty keen on a clear demarcation between science and religion, at least in so far as their work in the lab or on the computer goes (Figure 6.3).

In this, scientists generally adhere to a methodological distinction between the sphere of science and the sphere of religion, even though scientists with a religious faith may see science as falling entirely within the compass of God. Of course, many scientists, whether or not they have a religious faith, are concerned about the ethical implications of their work and may choose not to work in a particular area of science if they consider its practices to be ones with which they disagree—for example, research that causes animal suffering or makes use of material from elective abortions. But when it comes to trying to discern how the Universe runs, scientists with a religious faith work in the same way as those without it.

Returning to the issue of purpose in the Universe, the past twenty years or so has seen a number of top physicists, including some who do not have a conventional religious faith, more comfortable talking about the possibility that the Universe has a purpose. The reason is all to do with the anthropic principle.

The anthropic principle is understood in a number of ways but fundamentally it is the notion that any observations we make about the Universe are constrained by the logical requirement that these observations are made by us.[5] That doesn't sound particularly helpful but consider the question as to the age of the Universe. We said above that it's probably about 13.8 billion years old. Now let's make the assumption that we (i.e. the entities making the observations and calculations that allow us to arrive at this figure) are carbon-based. Well, physicists know a great deal about how long it takes to get a decent amount of carbon in

"I THINK YOU SHOULD BE MORE
EXPLICIT HERE IN STEP TWO."

Figure 6.3 Even scientists with a strong personal faith tend to keep it methodologically separate from their day-to-day professional ways of working.

the Universe. For a start, you have to have giant or supergiant stars
(smaller ones won't do). Then, for the carbon to be spread out in space
you need to have stars exploding as supernova. All this takes time—
which means that for us to be here to see it the Universe simply could
not be, for example, an order of magnitude younger.

   The anthropic principle can usefully be divided into two types: the
weak anthropic principle, which is widely accepted, and the much
more wacky-sounding strong anthropic principle, which is not.

The weak anthropic principle focuses on the fact that as physicists have learnt more and more about the Universe, it is clear that the values of what are often termed the fundamental physical constants—things like the charge on the electron, the gravitational constant, and the speed of light—need to be remarkably close to their actual values if carbon-based life (and that's the only life we know of) is to have evolved. It is difficult to quantify 'remarkably' but a conservative estimate across the twenty or so fundamental physical constants suggests that collectively this is a massive understatement.[6] A departure by much less than just one part in a million, million, million, million would be more than enough to prohibit the evolution of even the most rudimentary carbon life forms, let alone physicists capable of filling out grant applications.

What can we conclude from this fine-tuning, as it is sometimes called?[7] As one would expect (back to Figure 6.3), most physicists have not concluded that the Universe is therefore designed by some divine being or set of beings. At its simplest it might be that there is only one universe and we are very lucky (if you feel positive about the worth of life) that it just happens to be one in which intelligent life could evolve. Another possibility is that the twenty or so fundamental physical constants are not independent of each other but connected in some way that physicists haven't (yet) worked out so that the apparent fortuity of life is not quite so fortuitous. Another possibility is to take the anthropic principle at face value and keep on insisting that all this talk of the fortuity of life is to have failed to take seriously the point that intelligent life is here and without it we wouldn't be having this discussion so it's a logical fallacy to try to calculate the probability of life evolving without taking this into account—and if we take it into account the probability is 1 (i.e. certain). This argument suggests that it's a bit like trying to calculate the probability of a fee-paying spectator at a County Championship game at the Oval being keen on cricket—it is only people who are keen on cricket who pay to see County Championship games at the Oval.

Another possibility—the multiverse idea we mentioned in Chapter 5—is that there are actually lots of what we call 'universes' in the sense that communication between them is (probably) not possible and that these universes may differ in their fundamental physical constants. Perhaps the values of these are not fixed until soon after the origins of each universe in much the same way as our facial features are not entirely determined by our genes but are also affected by our development. In this case there might be a myriad of universes, only a relatively small number of which—possibly only one—allow physicists and the rest of us to evolve.

Finally, we turn to the strong anthropic principle—which was distinguished from the weak anthropic principle by Brandon Carter who first named the anthropic principle back in 1973 at a symposium held in Kraków in honour of Copernicus' 500th birthday. For Carter, the strong anthropic principle asserts that the Universe must be such as to admit the existence of observers within it at some stage. Others have built on this idea but the key point is that whereas in the weak anthropic principle there is a direction to the Universe (we end up with intelligent observers), but no purpose, in some versions of the strong anthropic principle, the Universe has a purpose—namely to give rise to intelligent observers.

The two of us are content with a conventional theological view of the Universe in which it is created by God and God's ends (purposes) include the provision of conditions that allow sentient beings to arise. We do not find persuasive the reverse argument, namely that the suitability of the laws of physics as we find them, combined with the apparent inevitability of evolution through natural selection, provides evidence that God exists. There are too many problems with such an argument to address them adequately here.[8] Charles Darwin cautioned against speculation when trying to work out why the Universe is as it is. As he put it, in a quotation that is often shortened:

> With respect to the theological view of the question;[9] this is always painful to me.—I am bewildered.—I had no intention to write atheistically. But I own that I cannot see, as plainly as others do, & as I sh[d] wish to do, evidence of design & beneficence on all sides of us. There seems to me too much misery in the world. I cannot persuade myself that a beneficent & omnipotent God would have designedly created the Ichneumonidæ with the express intention of their feeding within the living bodies of caterpillars, or that a cat should play with mice. Not believing this, I see no necessity in the belief that the eye was expressly designed. On the other hand I cannot anyhow be contented to view this wonderful universe & especially the nature of man, & to conclude that everything is the result of brute force. I am inclined to look at everything as resulting from designed laws, with the details, whether good or bad, left to the working out of what we may call chance. Not that this notion *at all* satisfies me. I feel most deeply that the whole subject is too profound for the human intellect. A dog might as well speculate on the mind of Newton.—Let each man hope & believe what he can.—[10]

Although Darwin mentions the Universe, he is primarily focused on evidence of design in living systems.

The purpose of life

When one of us used to teach biology to 16–19 year olds, he would sometimes ask his students what giraffes have that no other animals have. Almost inevitably, the first answer would be 'long necks' and almost inevitably no one would give the answer he was looking for: *baby giraffes*. In Aristotelian language, the final cause of a giraffe is to produce more giraffes. As we wrote in Chapter 1, to a reductionist evolutionist, the purpose of life is to produce more life, life that is as closely related to itself and as well adapted for survival and reproduction as possible. It is a truism that despite the immeasurable number of organisms that die each year without leaving any descendants, every organism alive today is the result of an unbroken chain of life, a chain that goes back to the origins of life. LUCA (the Last Universal Common Ancestor) is currently thought to have lived some 4.4 billion years ago, remarkably soon (150 million years or so) after today's Earth came into existence as a result of the planet Theia crashing into the then Earth, an event which sterilized Earth and led to the formation of the Moon.[11]

In everyday language, even the simplest organisms seem to have a sense of purpose. One can watch unicellular organisms under the microscope and see them moving towards the light if they are photo-synthetic, and pursuing prey if they are carnivorous.[12] Multicellular organisms often show a remarkable ability single-mindedly to pursue their aims as nature programmes on television illustrate. When it comes to birds and non-human mammals this can reach wonderful levels—the ways in which nests are built, food obtained, young provisioned and cared for, and predators avoided.

As the historian and philosopher of biology Michael Ruse points out, Aristotle (again!) was clear that such purposefulness did not necessarily require conscious thought.[13] As Aristotle put it:

> This is most obvious in the animals other than man: they make things neither by art nor after inquiry or deliberation. That is why people wonder whether it is by intelligence or some other faculty that these creatures work,—spiders, ants and the like. By gradual advance in this direction we come to see clearly that in plants too that is produced which is conducive to the end—leaves e.g. grow to make shade for the fruit. If

then it is both by nature and for an end that the swallow makes its nest
and the spider its web, and plants grow leaves for the sake of the fruit and
send their roots down (not up) for the sake of nourishment, it is plain
that this kind of cause is operative in things which come to be and are by
nature.[14]

Unicellular organisms, plants, fungi, and most animal species are not
capable of acting deliberately—they simply haven't got the necessary
brain power. It is always difficult to be certain when extrapolating
from humans to other species but there seem to be two errors to be
avoided when comparing us with other species with regards to our
deliberations. One is to assume that any organism that acts with
apparent intentionality is aware of what it is doing (we are back to our
point about watching unicellular organisms under the microscope).
The other is to assume that humans are unique in being able to act
deliberately.

Biologists always get nervous about humans being put in a completely different category to all other species. Consider language—it's
fine to see human language as immeasurably superior to the commu-
nicative abilities of other species but it is not wholly distinct. In the
same way, no one thinks that elephants, chimpanzees, dolphins, cor-
vids (crows and ravens), pigs, cats, and certain other mammals or birds
are capable of acting with the same degree of conscious thought and
planning that we see in humans but this does not mean that they lack
any such ability (Figure 6.4). Here is Jane Goodall writing about a young
male chimpanzee, Flint, his older brother Figan, his sister Fifi, and their
mother Flo:

> One day, when Flo was fishing for termites, it became obvious that Figan
> and Fifi, who had been eating termites at the same heap, were getting
> restless and wanted to go. But old Flo, who had already fished for two
> hours, and who was herself only getting about two termites every five
> minutes, showed no signs of stopping. Being an old female, it was pos-
> sible that she might continue for another hour at least. Several times
> Figan had set off resolutely along the track leading to the stream, but on
> each occasion, after repeatedly looking back at Flo, he had given up and
> returned to wait for his mother.
>
> Flint, too young to mind where he was, pottered about on the heap,
> occasionally dabbling at a termite. Suddenly Figan got up again and this
> time approached Flint. Adopting the posture of a mother who signals
> her infant to climb on to her back, Figan bent one leg and reached back

Figure 6.4 Chimpanzees show behaviours that strongly suggest they are self-aware and intentional.

his hand to Flint, uttering a soft pleading whimper. Flint tottered up to him at once, and Figan, still whimpering, put his hand under Flint and gently pushed him on his back. Once Flint was safely aboard, Figan, with another quick glance at Flo, set off rapidly along the track. A moment later Flo discarded her tool and followed.[15]

You might think that this is rather anecdotal—and Goodall was criticized at the time for this, though animal behaviourists are now more comfortable with this way of writing than many at the time were. But however Goodall's account is read, it suggests the same sort of self-awareness and intentionality that most of us are likely to find reminiscent of three- or four-year-old humans. One cannot rule out the solipsistic position that only one's own mind is sure to exist and therefore that one cannot be certain that even the external world is real, let alone populated by other humans with minds capable of thinking about such issues,[16] but the doctrine of solipsism is anathema to natural scientists in general and evolutionary biologists in particular. There is no consensus among scientists on possible distinctions between consciousness, awareness, self-awareness, and other terms and on how one might decide whether another species manifests these. Setting aside what some would consider to be armchair philosophizing in the pejorative sense, chimpanzees are examples of non-humans that seem purposeful in the fullest sense of the term.

Helping behaviour in non-humans

Before we focus in the remainder of this chapter on humans, it is worth considering helping behaviour in other organisms for two reasons. For one thing, some species seem at their most purposeful when helping other individuals. For another, as we mentioned in Chapter 3, at least some categories of helping behaviour seem to contradict the line of argument we have advanced above, namely that the purpose of life (at least for non-humans) is to leave as many viable offspring as possible. There are many examples where organisms help one another in ways that go beyond straightforward instances of parents bringing up their offspring. Darwin himself pondered the evolution of sterility in the social insects (ants, bees, termites, and wasps). In such species, many individuals never reproduce—indeed, they never seem even to try to reproduce (Figure 6.5). Instead, they serve the colony as a whole, typically

Figure 6.5 Honeypot ants are species of ants that have workers (called 'repletes' or 'rotunds') that function as living storage vessels. In common with other worker ants, they generally do not attempt to reproduce.

by assisting others in the colony to reproduce, either directly (for example, by provisioning them with food) or indirectly (for example, by protecting the colony from attacks). Darwin realized that what such individuals are in effect doing is to reproduce via others in their colony.

Nowadays we realize that the story is a bit more complicated—there are evolutionary battles within a colony as the various individuals do not all share identical interests—but the fundamental insight of Darwin holds good. This type of activity is nowadays named 'kin selection' as individuals are, effectively, reproducing via their kin (e.g. their siblings) rather than directly. Kin selection and reciprocal altruism, when one individual helps another individual (who may not even be in the same species) with the expectation (though this is not to imply any conscious awareness) that the time will come when such behaviour will be reciprocated and the altruist thus paid back, are important drivers of helping behaviour in non-humans.[17]

While the story of helping behaviour in non-humans is fascinating to biologists, it doesn't fundamentally alter the narrative that for non-humans the purpose of life—the end of life, the ultimate aim of life, without implying that this is intentional—is indeed, as far as mainstream evolutionary biology is concerned, to produce further copies of itself. Most people live as though there were more to human flourishing than that. Parents who simply want replicas of themselves seldom have happy children.

On being human

Being human (setting aside young children and those who are mentally incapacitated for whatever reason) involves making choices about matters for which we are responsible and that require decisions to be made. This element of agency is integral to human action—and the other side of our agency is that we have responsibility; a thermostat is a responsive agent—but not a responsible agent. Our agency is such that, unlike non-humans, we can create purposes for ourselves. One striking example of this, in a way that conflicts with the purpose of life for our non-human ancestors and relatives, is our use of contraception.

Contraception

A contraceptive is a method of preventing pregnancy (barrier contraceptives also function to reduce or prevent the transmission of sexually transmitted infections). Contraception has existed since ancient times but it is only in the last hundred years or so that it has become highly effective, widely available, and less subject to social stigma. The purpose of contraception is clear—but what is striking is how widespread contraceptive use has become, despite the fact that having children is, from the perspective of an evolutionary biologist, pretty much *the* purpose of life for humans. The probable reason why contraception can be so widespread, even though it runs contrary to the workings of evolution, is that evolutionary force of natural selection has had far longer to work on our sexual behaviour than on our use of contraceptives. Indeed, it is much harder for most people in the prime of life to abstain from sexual intercourse for long periods of time (say, several years) than it is to use contraception for the same period of time.

Contraception demonstrates our human capacity to manipulate our purposes. We decide that we don't want any more (or any) children, at least not now, but that we don't want to abstain from sexual intercourse, and so we choose to use contraception. We are the only species to do this. At one time, the use of contraception was frequently viewed as a selfish act—on the grounds that it allowed a heterosexual couple to engage in sexual intercourse without fear that a pregnancy might result. Nowadays, though, most people view contraception more positively, as a way of exercising control over when children might be born and managing the size of families. Added to that, we are more aware nowadays that humans are placing an increasingly unsustainable load on the Earth's natural resources. Leaving aside the colonization of space, the only realistic ways out of this are to increase the Earth's carrying capacity, to lower our resource demands, or to limit population size. At present, the last of these three options, though we aren't there yet, seems a better bet than either of the first two, particularly given the huge and near worldwide decreases in infant mortality rates over the past century.

How necessary are children if humans are to feel that life is purposeful? Albert Einstein would often use a thought experiment (in German *Gedankenexperiment*) as a tool for clarifying his own thinking and communicating his concepts to others. In a similar way, novelists can use stories to explore how humans would behave in fictitious circumstances. Sometimes this is rather like a scientist taking out a specified factor, such as removing the effect of air by using a vacuum or removing thermal agitation by using a cryostat. In just such a way Jane

Austen largely removes the Napoleonic wars in her novels. Sometimes a writer creates a situation which could not actually be realized for practical or ethical reasons. In the *Lord of the Flies*, William Golding maroons two groups of well-educated adolescent schoolboys on a Pacific island and runs a thought experiment to explore how they would behave. Before long their incipient attempts at civilization descend into savagery.

In *The Children of Men*, the detective novelist P. D. James conducts a thought experiment which would be even more impossible in practice. What would happen if there were to be no more children—ever? How then would people live? What would they live for? The story opens in Oxford with the university authorities trying to decide whether to resurface the stonework of the Sheldonian Theatre, designed by Christopher Wren in the seventeenth century. If before long there would be no one to see it, why go to all the bother and expense? Two central characters in the book are cousins. One of them, the narrator, is an academic, who finds purpose in helping a group of five people who are trying to alleviate some of the immediate problems. In the second part of the book he finds a deeper purpose, though we won't spoil the story for you. His cousin Xan is a politician, who finds purpose in keeping everything orderly. He ends up running the country as Warden of England. The last generation to be born, called the Omega children, run amok. By removing such a basic aspect of life as nurturing and educating the next generation, this thought experiment of no more children stimulates the reader to ask profound questions about their own lives. What are we living for? What is our purpose?

Helping behaviour

Humans are a social species. We live in communities, we help one another, we take from others, and we receive help from others. As we discussed in Chapter 3, the same evolutionary forces that help explain why non-humans cooperate and help one another also apply to us. Much of our helping behaviour can be explained by kin selection—when those we help are related to us so that, in a sense, we can reproduce via them as well as directly ourselves—and reciprocal altruism—when we help someone else (who may or may not be related to us), with the expectation that they will subsequently repay us. Indeed, for most of us, this works the other way too. If someone helps us, we remember and are more likely subsequently to help them.

There's a rather nice example that follows of a behaviour as a predicted consequence of reciprocal altruism in humans that was discovered by Melissa Bateson and colleagues.[18] The study was conducted in a naturalistic (i.e., non-laboratory) setting on 48 members of the Division of Psychology at the University of Newcastle (there are no prizes for guessing where Bateson and her co-authors work). The university coffee room operates an honesty box where people are supposed to put their money if they make themselves a tea or coffee using the tea, coffee, and milk provided. A black-and-white A5 notice states the prices. This system has been in operation for years. What the experimenters did was to add a banner to the notice. One week the banner featured a pair of human eyes, the next week some flowers. The experiment ran for a total of ten weeks and a new banner was used each week.

The results of the study are shown in Figure 6.6. In the weeks when the banner featured an image of human eyes, almost three times as much money was collected compared to when it featured flowers. None of the participants when questioned afterwards had realized that an experiment was being undertaken. The interpretation of the findings is that when, subconsciously, we think that someone is watching us, we are more likely to be honest—and to reciprocate. When we don't presume that someone is watching us, we are more likely to be dishonest, to cheat. In an apocryphal not dissimilar experiment, a fixed-price meal in a cafeteria allowed for one roll, one main course, and one piece of fruit per person. When it was discovered that some people were taking more than one roll, a sign was put up over the rolls, 'Please do not take more than one roll—God is watching you!' Before long another sign appeared over the fruit, 'Take as many pieces of fruit as you like—God is checking the rolls!'

Bateson's study illustrates that much of our behaviour is undertaken without us being aware of the influences on our actions. Nevertheless, and this is key to human flourishing, we are capable of reflecting on our actions, even if we all too often fall short of what we hope we would do. As Paul put it, 'For I do not do the good I want, but the evil I do not want is what I do.'[19] More positively, we are rational creatures (though we will examine in Chapter 8 the extent to which models of human behaviour have sometimes overestimated the extent to which this is the case) and this has profound consequences for our actions, including our helping behaviour.

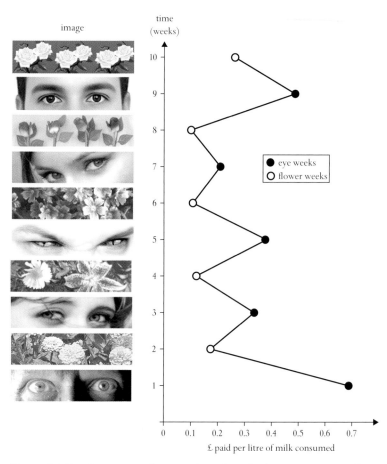

Figure 6.6 People are more honest when they think someone is watching them. In an experiment, the amount paid into an honesty box for coffee, tea, and milk at a university depended on whether the banner displayed on the notice that gave the prices (50p for coffee, 30p for tea, 10p for milk) showed human eyes or flowers.

In his book *The Expanding Circle*, the moral philosopher Peter Singer argues that altruism began as a drive to protect one's kin and those in one's community but has developed over time into a consciously chosen ethic with an expanding circle of moral concern.[20] We see this as children develop their moral sense and move from egotistical behaviours to some sort of internalized set of ethical principles. The key point is that for all that we naturally favour our immediate family, more distant relatives, and those around us, we have a natural (i.e. evolved) tendency to empathize with anyone in distress or need *and* the rational parts of our minds tend to tell us that it's only fair to help if we can when others are in need.

Of course, there are always exceptions. Psychopaths are often said to lack empathy. It may be more accurate to say that they can turn it on or off more or less at will. Most of us if we see someone in pain can't help empathizing. In one study, criminal psychopaths were asked to watch videos of one person hurting another and the area of the brain that responds to pain was monitored.[21] Only when the participants were asked to imagine how the person in the video who was being hurt felt did the areas of their brain that respond to pain light up in the same way as controls (i.e. people who were not criminal psychopaths). The researchers suggested that this capacity of psychopaths to control when they empathize might explain why they can be callous at one time and charming at another.

For most of us (presumably not for most psychopaths), giving to others tends to make us happier. In one study, university students (a lot of psychology studies are undertaken on university students) were given US$5 a day for five days and required to spend it on the same thing for all five days. Half the students (chosen randomly) were told to spend the money on themselves; the other half were told to give it away—as tips or in an online donation to a charity. The students who kept the money for themselves ended up less happy than the students who gave it away.[22]

Religion and a sense of purpose

The Westminster Shorter Catechism is a summary of what was then considered essential Christian doctrine written in 1646–7 by a synod of English and Scottish theologians and laity. It was intended to bring the Church of England into greater conformity with the Church of Scotland and was used to educate children and others 'of weaker capacity'. The first of its 107 questions is 'What is the chief end of man?', to which the model answer is 'Man's chief end is to glorify God, and to enjoy him forever'. The language may sound somewhat quaint to contemporary ears but the answer makes clear a fundamental point—namely, that someone with a firm religious conviction is likely to see the (more modestly, 'a') purpose of their life as relating to God and responding appropriately to what they believe God is calling them to do.

In the world's scriptures, there are repeated instances of God calling to people. If we restrict ourselves to the Jewish scriptures, *bat ḳōl* (בַּת קוֹל) is 'the voice of God' and God calls his people from the book of Genesis through to the last of the prophets—Haggai, Zechariah, and Malachi. Famous examples include Moses' call from out of the burning bush, Elijah hearing God not in the wind, the earthquake, or the fire but in a still small voice, and Ezekiel hearing God tell him to prophesy to dry bones so that they became clothed with tendons and flesh and skin (Figure 6.7).

Few of us claim to hear God or discern God's purpose for our lives with the clarity of the Jewish prophets, and we are beginning to use 'purpose' in a sense that is very close to that of 'meaning', to which we will turn in our next chapter. Is it unsurprising that people for whom religion is important are more likely to feel that life has a purpose (or meaning) than those for whom religion is not important? A study that drew on data from 84 countries found that in response to the question 'Do you feel your life has an important meaning or purpose?',

Figure 6.7   Part of the Ezekiel Panel in the Synagogue of Dura Europos, Syria and dating from before 244 CE. The prophet Ezekiel is shown three times—the left-hand representation corresponds to Ezekiel 37: 1: 'The hand of the Lord was upon me, and he brought me out by the Spirit of the Lord, and set me down in the midst of the valley.'

92 per cent of those who said they had a religious affiliation answered 'yes' and 6 per cent 'no', whereas the corresponding figures for those who identified as secular, nonreligious, atheist, or agnostic were 83 per cent and 14 per cent.[23] This difference held for those whose religion was Buddhism, Christianity, Hinduism, Islam, Judaism, and 'Other'. Michael remembers debating once with Andrew Copson, Chief Executive of Humanists UK. The point at which he felt that many in the audience suddenly lost confidence in Copson's arguments was when Copson asserted that life was meaningless.

In the film *Silent Running*, the botanist Lowell found purpose in being prepared to lay down his life, not for his friends but for a greenhouse of plants and non-human animals. Our examination of religion and a sense of purpose has suggested that there is an important distinction between whether one feels life has one or more objective purposes to which one is called to respond or whether one decides on one's own purposes in life. This distinction is perhaps even more important when it comes to finding or creating meaning in one's life.

Purpose and meaning

In the context of human flourishing, purpose and meaning are sometimes used almost interchangeably, so that a life of purpose might be spoken of as meaningful. We have used *purpose* in the specific sense of the reason for which something is done or made, or for which it exists.[24] This then frees us to use *meaning* for other senses, such as significance and underlying truth.[25]

In Chapter 1, we illustrated the distinction between purpose and meaning by considering the notes Charles Darwin made in July 1838 while he was pondering whether to propose to his first cousin, Emma Wedgewood—which he eventually did (on 11 November 1838). In his typical, methodical way, Darwin listed both advantages and disadvantages to marriage, noting, in its favour, amongst other arguments:

> Children... Constant companion, (& friend in old age) who will feel interested in one,—object to be beloved & played with... Home, & some- one to take care of house—Charms of music & female chit-chat.—These things good for one's health... a nice soft wife on a sofa...[26]

It may not be the most romantic passage of all time but it caused Darwin to conclude that it 'proved necessary to Marry'. And Darwin has covered what many would regard as the main purposes of a marriage—life-long companionship with one person and an arrangement within which children can be conceived and brought up. He even mentions the benefits of marriage for one's health, which we considered in Chapter 3.

Meanings are, in a sense, richer than purposes. We can talk about a carburettor having a purpose: before the advent of fuel injection systems, they mixed air and fuel in the right proportions for internal combustion engines. However, it sounds odd to talk about the meaning of a carburettor—and we are confident that this is not a question on which carburettors themselves dwell.

We often hear about people trying to find purpose in life when meaning is what is meant. Someone may fulfil a whole series of purposes—getting enough to eat, keeping healthy, maintaining somewhere to live, ensuring they have enough money to get by—and still feel that their life lacks meaning. In terms of the three dimensions to human flourishing that we covered in Part I, a person's material needs may be satisfied, they may enjoy close relationships, and they may even pay attention to the transcendent. And yet they may still find that there is something their life lacks that could make it more meaningful.

Notes

1 Falcon, A. (2019) Aristotle on causality, *The Stanford Encyclopedia of Philosophy*, 7 March. Available at https://plato.stanford.edu/archives/spr2019/entries/aristotle-causality.

2 Barrow, J. D. (2011) *The Book of Universes: Exploring the Limits of the Cosmos*, New York: W. W. Norton.

3 Eddington, A. S. (1928) *The Nature of the Physical World: The Gifford Lectures 1927*, Cambridge, UK: Cambridge University Press, pp. 56 7.

4 The way in which ordered life needs to create more disorder elsewhere was famously discussed by Erwin Schrödinger in his 1944 book *What is Life?*, Cambridge, UK: Cambridge University Press.

5 Bostrom, N. (2002) *Anthropic Bias: Observation Selection Effects in Science and Philosophy*, New York: Routledge.

6 Rees, M. (1999) *Just Six Numbers: The Deep Forces that Shape the Universe*, London: Weidenfeld & Nicolson.

7 Paul Davies calls it the Goldilocks Enigma after the story of the three bears in which Goldilocks eventually finds porridge of the right temperature, a chair of the right size, and a bed that is neither too hard nor too soft. See

Davies, P. (2006) *The Goldilocks Enigma: Why Is the Universe Just Right for Life?*, London: Allen Lane.

8 Briggs, G. A. D., Halvorson, H. and Steane, A. M. (2018) *It Keeps Me Seeking: The Invitation from Science, Philosophy, and Religion*, Oxford: Oxford University Press, pp. 135–74.

9 This is the question of God's existence—Darwin is writing a letter six months after the publication of *On the Origin of Species*.

10 Letter to Asa Gray, 22 May 1860, Darwin Correspondence Project, Letter 2814. Available at https://www.darwinproject.ac.uk/letter/DCP-LETT-2814.xml.

11 Betts, H. C., Puttick, M. N., Clark, J. W., Williams, T. A., Donoghue, P. C. J. and Davide, P. (2018) Integrated genomic and fossil evidence illuminates life's early evolution and eukaryote origin, *Nature Ecology & Evolution* **2**, 1556–62. For a discussion of rival theories of the origin of the Moon see https://royalsociety.org/science-events-and-lectures/2013/origin-moon/.

12 Numerous short videos can be found on the Internet by searching for organisms like *Euglena* or *Amoeba*.

13 Ruse, M. (2018) *On Purpose*, Princeton, NJ: Princeton University Press.

14 Aristotle (1984) *Physics*, in *Complete Works of Aristotle*, vol. 1, ed. J. Barnes, Oxford: Oxford University Press, p. 340. Ignore what he writes about leaves growing to make shade for fruit.

15 Van Lawick-Goodall, J. (1971) *In the Shadow of Man*, London: Collins, pp. 114–15.

16 One widespread approach to attempting to determine whether an animal is self-aware is the mirror test. In essence, the animal is marked—e.g. by the application of rouge—without it noticing (e.g. because it is asleep) on a part of its body that it cannot normally see. When the animal is subsequently awake, it is given access to a large mirror. If it investigates the mark, this is taken as evidence that the animal 'presumes' the reflected image shows itself and, in this sense, is self-aware. Chimpanzees are one of the few species that pass the test.

17 Nowak, M. A. and Coakley, S. (Eds) (2013) *Evolution, Games, and God: The Principle of Cooperation*, Cambridge MA: Harvard University Press.

18 Bateson, M., Nettle, D. and Roberts, G. (2006) Cues of being watched enhance cooperation in a real-world setting, *Biology Letters* **2**(3), 412–14.

19 Romans 7: 19.

20 Singer, P. (1981) *The Expanding Circle: Ethics, Evolution, and Moral Progress*, Princeton, NJ: Princeton University Press.

21 Meffert, H., Gazzola, V., den Boer, J. A., Bartels, A. A. J. and Keysers, C. (2013) Reduced spontaneous but relatively normal deliberate vicarious representations in psychopathy, *Brain* **136**(8), 2550–62.

22 O'Brien, E. and Kassirer, S. (2018) People are slow to adapt to the warm glow of giving, *Psychological Science* **30**(20), 193–204.

23 Crabtree, S. and Pelham, B. (2008) The complex relationship between religion and purpose. Gallup, 24 December. Available at https://news.gallup.com/poll/113575/complex-relationship-between-religion-purpose.aspx.

24 According to the Oxford *English Dictionary*.

25 According to the Oxford *English Dictionary*.

26 Darwin on marriage, Darwin Correspondence Project, Cambridge University Library, DAR 210.8:2. Available at https://www.darwinproject.ac.uk/tags/about-darwin/family-life/darwin-marriage.

7

Meaning

Someone whose consistently feels that their life is meaningless will find it hard to say that they are flourishing. But people can find meaning in very different ways. Some people find it in the specifics of daily life—time with friends, going to a great concert, the satisfaction from cooking a successful meal, travel; each of these can bring meaning and contribute to answering the question 'where is meaning in *my* life?' Others search for an overarching framework within which to find an answer (or answers) to the rather larger question 'what is *the* meaning of life?'

Behind these questions lie two deeper questions. First, is meaning something to be discovered, perhaps like truth, or is it something to be chosen? Or is it a subtle combination, so that we all, individually and in community, are tasked with creating meaning within the external constraints of reality and the internal constraints of our abilities and desires? Second, is meaning an impersonal thing, like gravity, or does it have to be imparted by an agent, like love?

In this chapter we will use our three dimensions of human flourishing—the material dimension, the relational dimension, and the transcendent dimension—as a way of examining what meaning is and how people find meaning. Philosophy, theology, and psychology each have a great deal to say about what is meaningful and we will draw on all three of these disciplines. We start with meaning in communication.

What does a message mean?

Henry Wadsworth Longfellow, with a certain poetic licence regarding the historical facts, immortalized what may have been the most compressed signal in the history of Boston, Massachusetts:

> Listen, my children, and you shall hear
> Of the midnight ride of Paul Revere,
> On the eighteenth of April, in Seventy-Five:

Hardly a man is now alive
Who remembers that famous day and year.

He said to his friend, 'If the British march
By land or sea from the town to-night,
Hang a lantern aloft in the belfry-arch
Of the North-Church-tower, as a signal-light,—
One if by land, and two if by sea;
And I on the opposite shore will be,
Ready to ride and spread the alarm
Through every Middlesex village and farm,
For the country-folk to be up and to arm.'[1]

The message was actually for Charlestown, not for Paul Revere, since he first instructed the sexton Robert Newman to set two lanterns and then crossed to the opposite shore, but let's not spoil a good story. The point is that the signal contained only one bit of information. In binary notation, we might say 0 (one lantern) meant that the British would attach by land, and 1 (two lanterns) meant that they would attack by sea. That was all the people of Charlestown needed to know.

Information theory as we now have it was the brainchild of Claude Shannon. In 1948 he published a paper entitled 'A Mathematical Theory of Communication'. The following year it was republished as a book, in whose title 'A' was changed to 'The', to reflect the generality of the theory. Crucial to his theory was the amount of uncertainty on the part of the recipient of the information. In Newman's message, the only uncertainty was whether the British would arrive by land or by sea, thus requiring one bit to provide the information. If the recipients had not known which night, or which army, or even whether there would be an attack, one or two lanterns would not have sufficed.

Shannon's mathematical theory was concerned solely with technical information. It did not cover meaning, what is sometimes call semantic information. Newman's signal transmitted the information, 'whether by land or by sea'. It gave no indication of the truthfulness of the information; for that the recipients had to rely on the history and integrity of Robert Newman (and hope that the number of lanterns had not been set by an enemy by way of fake news). Nor did the communication tell the recipients what to do about the information; that required prior knowledge from discussions which had taken place in preparation. The meaning of the information depended on the recipients.

Meaning depends on the preparedness of the recipient in contexts other than military communications. A will may specify that all or part of an estate is to be divided in equal shares *per stirpes*. A lawyer knows exactly what that means; someone who has never administered a will might not. The meaning of legal documents generally depends on a great deal of prior knowledge. One of us has just shown our students how beautifully compact tensors are for describing anisotropic properties of materials, but it takes more than one lecture to explain the meaning of the mathematical notation.

What about meaning in the creative arts? In prose, poetry, painting, sculpture, song, music, dance? How should a dancer respond when asked what the dance means, other than to reply that if she could have told it in words she would not have danced it? To find meaning in these expressions demands not only preparation on the part of the recipient, but also response. The poet T. S. Eliot wrote in a letter to Claude Collier Abbot:

> I am pleased that you like *The Waste Land* and wish that I could tell you more about it. It is not evasion, but merely the truth, to say that I think in these cases that an explanation by the author is of no more value than one by anybody else. You see, the only legitimate meaning of a poem is the meaning which it has for any reader, not a meaning which it has primarily for the author. The author means all sorts of things which concern nobody else but himself, in that he may be making use of his private experiences. But these private experiences are merely crude material, and as such of no interest whatever to the public.[2]

It may well be that the meaning for each reader is different, and therefore different from the meaning for the author. Nevertheless, the author, or composer, or painter creates something that allows them to impart what they are passionate about to the recipient. In *Evening in the Palace of Reason*, James Gaines describes the encounter between Bach, whose music was the outpouring of his worship of God, and Frederick the Great, who scorned religion, collected people like the worst of social media, and loved music for its melody. They met in Potsdam in May 1747, and almost immediately Frederick set Bach the challenge of composing—*ad lib*—a three-part fugue from a series of seventeen notes that were as ill-suited to counterpoint as it was possible to devise. By all accounts, Bach not only succeeded in dazzling the audience with his spontaneous improvisations, he even managed to weave in the entire theme twelve times in the space of seventeen minutes.

After his departure from the royal court, Bach developed what he had begun that evening into what he called *Musical Offering* (*Musicalisches Opfer*), which he dedicated (the German word can also mean consecrated) to Frederick. Did the music have a *meaning*? Gaines thinks so:

> All of the oddities contained in the work—the harrowing descent in *galant* passages, the melancholy fate of the king's fortune, the song to glory that goes nowhere, the German dedication, the Scriptural invocation to 'seek and find' God's mercy rather than the harsh, eternal judgement of God's own canon law, the setting of a church sonata—all of these were of a piece, and this is what they say: Beware the appearance of good fortune, Frederick, stand in awe of a fate more fearful than any of this world has to give, seek the glory that is beyond the glory of this fallen world, and know that there is a law higher than any king's which is never changing and by which you and every one of us will be judged. Of course that is what he said. He had been saying it all his life.[3]

It seems that if Frederick understood the message, he did not heed it. Gaines quotes his biographer William Reddaway: 'Through all his life—in his councils, in his despair, in his triumph, and in his death—Frederick, almost beyond parallel in the record of human history, was alone.'[4]

Communication takes place in the context of a relationship in which *information* is transferred from an *informer* to an *informee*, if we can use these words without attendant baggage. Sometimes the relationship is one way, as in a broadcast. To encourage interaction, some programmes invite audiences to send in responses. In other creative arts the relationship is also asymmetric, almost absolutely so when communications are disseminated through published or streamed media, though there can be asynchronous responses through reviews, and more interactively but still asymmetrically in live performances through laughter or applause.

Communication may be more fully reciprocal between two friends, whether they are enjoying frivolous banter or engaging in deep discussion. Meaning in conversation involves both intentionality on the part of the sender and interpretation on the part of the recipient. Through dialogue they may converge on a meaning, whether or not they reach agreement on the content. As Don Adriano de Armado asked Costard, they may ask, 'What meanest thou?'[5]

This process of clarifying meaning in conversation can be thought of as a Bayesian process, along the lines which we discussed in Chapter 5, although we do not recommend trying to put numbers into a formula in this example. You tell me something, perhaps how you feel about something which has happened. I have some sense of what you mean, but I may have a low level of confidence and a high level of uncertainty. So I ask you a question, or possibly make an observation, which I hope will elicit further information from you. In a healthy conversation your response will decrease my uncertainty and increase my confidence that I understand what you mean. I shall have improved my knowledge of the meaning which you intended to impart.

To what extent does meaning demand decisions on the part of the recipient? In his exploration of the drama, music, symbolism, and philosophy of Wagner's Ring Cycle, Roger Scruton explains how he chooses his interpretation of the operas in order to elucidate their meaning about the search for meaning. 'In this chapter I give an interpretation of what the story is about. It is only one of many possible interpretations. But it will serve as a framework within which to place the philosophical and moral themes that I go on to address.' The chapter concludes that Wotan:

> is 'groping towards a tragic ending', one that will dignify his doomed government, endow it with a meaning, show it to be deeply worthwhile and not just a random intrusion into a world that has no design unless we impose one. At length, thanks to Brünnhilde, who ensures that by losing everything, she ends everything for the gods, Wotan finds the tragic consummation that he has unconsciously so often evaded. *The Ring* is the story of these two central characters, and of their search for meaning in a world where meaning exists only if we ourselves provide it.[6]

Scruton is not here endorsing Wotan's and Brünnhilde's beliefs, or even Wagner's, but he is highlighting an approach developed by Nietzsche and others which denies that any meaning exists beyond what humans devise. John Cottingham finds that too much. He refers to the earlier view of the Greek philosopher Protagoras, who, similarly, asserted that man is the measure of all things: of what is, that it is, and of what is not, that it is not:[7]

> By supposing the unaided human will can create meaning, that it can merely by its own resolute affirmation bypass the search for objectively sourced truth and value, he [Nietzsche] seems to risk coming close to the Protagorean fallacy. For meaning and worth cannot reside in raw will alone: they have to involve a fit between our decisions and beliefs and what grounds those decisions and beliefs. That grounding may, as some religious thinkers maintain, be divinely generated; or it may be based on

something else—for example certain fundamental facts about our social or biological nature. But it cannot be created by human fiat alone.[8]

Is it possible to find meaning as an impersonal reality, without that meaning having been imparted by a conscious agent? How do people look for meaning in our three dimensions of human flourishing?

The material dimension to meaning

Many in the affluent society feel they have too much 'stuff' and occasionally have clear-outs. But not the artist Andy Warhol. Over the last 13 years of his life until he died in 1987, he consigned 300,000 of his everyday possessions to cardboard boxes called 'Time Capsules'. Each of these 610 Time Capsules is crammed full of letters, magazines, newspapers, receipts, photographs, invitations, pencils, and other material.

Warhol was wealthy and had the space to keep all his possessions. Compulsive hoarders generally don't. A person with 'hoarding disorder' is either unwilling or unable to discard some of their acquired objects, to the extent that their home is so full that it causes significant problems (Figure 7.1). In the worst cases, such hoarding can be directly responsible for someone's death. Billie Jean James was a 67-year-old

Figure 7.1 The living room of a compulsive hoarder.

peace activist who lived in Las Vegas with her husband. The house was so full of her possessions, many bought at thrift stores, that the couple could only move along narrow paths constructed between the mountains of stuff. In 2010, Billie Jean James went missing. The police looked for her and even conducted a number of searches of the couple's home with dogs but could not find her. Four months after her disappearance, her husband was shocked to find her dead body under a pile of clutter. The police later explained that the failure of the dogs to find her body must have been the result of the general mess and all the other smells.[9]

It is easy to look at tragic cases like that of Billie Jean James and wonder how a desire for material possessions could so come to dominate a person's life. Even if it doesn't kill you, compulsive hoarding is associated with substantial health risks, general impairment to functioning, and (perhaps unsurprisingly) economic burdens.[10] It causes distress among family members and leads to arguments and relationship breakdowns. People who suffer from compulsive hoarding often, in other people's eyes, seem remarkably unwilling or unable to recognize the problem. In this they differ from most people with obsessive compulsive disorders (such as excessive hand washing) who are generally very aware of the problems that their obsessions cause them. Other obsessive-compulsive disorders can often be treated successfully with cognitive behavioural therapy. Compulsive hoarding generally cannot.

At the other end of the spectrum from hoarders are those who practise 'simple living', including those influenced by Marie Kondo.[11] As always, there are extremes. In 2001, the British artist Michael Landy stood in a former C&A shop in Oxford Street, London. In front of him in boxes were the 7,227 items that he had accumulated over the 37 years of his life. Over the course of a fortnight, each and every one of these items—clothes, letters, artworks, furniture, his car—was shredded, crushed, dismantled, or otherwise destroyed by Landy and his team of assistants. By the end, the only thing Landy possessed was the blue boiler suit he was wearing. Landy called the project—which can be seen as a piece of performance art—'Break Down'. Landy is happier to talk about his own work than the laconic Warhol was and subsequently said, 'It was the happiest two weeks of my life.'[12] He explained: 'In a sense, the message was: where are we heading? . . . The more stuff people have, the more successful we perceive them to be—but if we all end up with 7,227 things, then we won't have a planet.'

The art critic Alastair Sooke has suggested that Break Down functions as a contemporary *memento mori*: 'all of us, to differing degrees, use possessions to construct our identities and project ourselves to others—yet here was a man wilfully obliterating his material existence to the point of total annihilation.'[13] A memento mori (Latin for 'remember death') is an artistic or spiritual reminder of the inevitability of death. In Western culture, memento mori are often discrete reminders, such as the mourning rings or brooches that used to be worn in memory of someone close to you who had died. Mourning brooches might contain a small portion of the person's hair—or someone else's; in the mid-nineteenth century, 50 tons of human hair a year was being imported into England for use by jewellers.[14]

In other cultures, though, far more may be made of death. Among the best-known traditions is the Mexican *Día de Muertos* (Day of the Dead). Dating from pre-Colombian culture, in Aztec times it was celebrated for a whole month. Nowadays, the principal focus is on November 1st (in memory of children) and November 2nd (in memory of adults). How *Día de Muertos* is commemorated varies from place to place in Mexico but a common theme is one of celebration more than mourning. The reasoning is that it is believed that the dead would be insulted by sadness. The dead are seen as part of today's community, awakened on *Día de Muertos* from their eternal sleep to share in the celebrations with their loved ones.[15]

Remembering the finitude of life, whether through mourning brooches, *Día de Muertos*, or other customs, can help curb material acquisitiveness—at least in theory. Yet many of us derive pleasure from material objects without succumbing to excessive hoarding. Less extreme versions of the simple life than Michael Landy's come under the overall heading of 'simple living'. One of the most famous individuals who adopted a simple life was Mahatma Gandhi (Figure 7.2). Gandhi was born in a Hindu family in Gujarat in 1869, trained in law at the Inner Temple in London, and was called to the bar (i.e. qualified as a barrister) in 1891. He worked in South Africa for 21 years, returning to India in 1915. Here he began to organize non-violent protests against injustice and discrimination.[16] In 1921 he became leader of the Indian National Congress and in the same year began to live modestly in a self-sufficient residential community. He started to wear the tradition loincloth and, in winter, a shawl, both of which he made using yarn that he spun on a traditional Indian spinning wheel as a mark of

Figure 7.2 Mohandas Karamchand Gandhi (1869–1948), commonly known as Mahatma Gandhi, eschewed materiality and lived a simple life at his choosing.

identification with the rural poor. He ate simple vegetarian food and undertook long fasts both as a means of self-purification and for political protests.

Simple living isn't a single coherent movement; it is more of an umbrella term. Reasons for adopting such a life are many and include reducing one's impact on the environment, alleviating stress, seeking to become more self-sufficient for reasons of personal satisfaction, and living with fewer possessions for spiritual reasons. Central to it is the notion of choice; it isn't simple living to live with few possessions if one has no alternative; the homeless live simply, at least with regards to what they own, but they aren't practising simple living. Certain Christians who separated themselves from the world and lived simply became known as 'Plain people', such as the Amish, who were portrayed in the 1985 film *Witness*, which won two Oscars, and the Shakers. Furniture made by the Shakers is now so valuable that it is principally found in museums or collected by the very rich, which is ironic, given that it was made simply, without decoration or elaborate details, as 'an act of prayer'.

A less isolationist Christian group that practises plain living are the Quakers—characterized by their belief since their origins in the seventeenth century that there is 'that of God in everyone', that women as well as men can be ministers, that we can all experience Christ directly, and that slavery, participating in wars, and the swearing of oaths are all wrong. Partly because of their plain living—and consequent lack of acquisitiveness—they came to be known for their honesty and trustworthiness, which led to considerable business success; Barclays, Lloyds, Friends Provident, C. & J. Clark, Cadbury, Rowntree, and Fry were all Quaker companies.

It is perhaps unsurprising that a number of religious groups have advocated that we should rely less on material possessions. In *A Man for All Seasons*, Robert Bolt portrays Thomas Moore as a man of utmost sincerity and integrity who is uncompromising on matters of conscience. By contrast, Richard Rich is concerned only with the acquisition of wealth and position. After Rich commits perjury at Moore's trial, Moore asks about a chain of office which Rich is wearing, to be told that Sir Richard is now Attorney General for Wales. Alluding to a question asked by Jesus, 'For what shall it profit a man, if he shall gain the whole world, and lose his own soul?',[17] Moore retorts, 'For Wales? Why Richard, it profits a man nothing to give his soul for the whole world . . . But for Wales!'[18] Out of deference to our friends from Wales, we would never endorse a low value being put on that country, but we do affirm the force of Jesus' question. What about other aspects of material culture? What about works of art, for instance? There seems to be an important distinction between admiring a work of art and owning a work of art. The person who finds meaning in a work of art exclusively through their ownership of it fails to appreciate the work of art for what it is, a portrayal of meaning that transcends words.

Too great a desire to own art—or anything else—can lead some people to theft. Some art theft is rather like the compulsive hoarding we discussed above. Stephen Blumberg was arrested in 1990 and subsequently convicted of stealing about 20,000 books valued at a total of $5.3 million from 237 libraries and museums across the USA and Canada. At his trial Blumberg argued that the government was plotting to keep ordinary people from having access to rare books, and so his aim was to liberate and release such books in an attempt to thwart the government. This argument did not convince the jury. He was sentenced to 71 months in prison and a $200,000 fine.[19]

Much art theft is simply for the purposes of making money. But the most famous works of, once stolen, cannot be sold on the open market—though they can be sold on the black market or used for blackmail and other illegal purposes. Sometimes art is stolen (often to order) when the person ultimately responsible for the theft has no intention of selling it. The art is sold simply so that a 'collector' can enjoy it; it's all about acquisitiveness and possession. The largest ever art theft (in terms of financial value, an estimated $500 million) took place in 1990 at the Isabella Gardner Stewart Museum in the USA. Thirteen works of art were stolen including Vermeer's *The Concert* (Figure 7.3), the most valuable art work ever stolen. None of them has ever been recovered and it is possible that they have ended up with one or more collectors.

Figure 7.3 Vermeer's *The Concert*, the most valuable stolen piece of art work in the world.

Many a family has fallen out over the distribution of heirlooms from a will. Often the cash value of the items is small compared with, say, a deceased parent's house or other assets. But the emotional value can be large enough to tear siblings apart. Because the objects have been in the family for many years, especially through the formative years of childhood, they can be invested with meaning that vastly exceeds their material content.

The relational dimension to meaning

Objects can be a source of great pleasure. They can help bring meaning to our lives and contribute to our flourishing. But for most of us, once our basic material needs have been met, while objects can provide pleasure and be meaningful for us, it is relationships that are more important—and often longer-lasting. Of course, as with the hoarding of newspapers, rubbish, books, or other objects, relationships can be taken to excess. This is true even with respect to pets. There are various daytime TV programmes (extensive research was undertaken for this book) that show how some people can end up drowning in pets or rescue animals. What started off as a handful of cats, dogs, hedgehogs, or whatever soon becomes so many that they take over a person's life—benefitting neither the pets nor the person. But for every person who ends up with too many pets, there are many, many who not only look after their pets well but derive meaning from their relationships with them.

Relationships with people too can, of course, be taken to excess. Of the kind of person who 'lives for others', it is sometimes said that you can tell the 'others' by the hunted looks on their faces. What has been called 'obsessive love disorder' speaks for itself. The affected person fixates on one other person, is excessively attracted to them to a degree that is not reciprocated, and becomes extremely possessive and jealous. Stalking can result.

But all this is to talk of excesses. Many of us, perhaps most of us, derive great worth from our relationships with other people and such relationships help us to find meaning in our lives.[20] This is common sense and the stuff of novels, poetry, and plays. It is also backed up by empirical research. For example, in one piece of research, a series of related studies were undertaken in which young adults were asked about the most important source for them of perceived meaning.[21] The researchers found that 68 per cent of their sample identified family as their primary source of meaning and another 14 per cent friends, so that for 82 per cent of participants, personal relationships were the primary source of meaning.

Other studies have looked at what happens to people's views as to the meaning of their lives when relationships go awry. There is a limit (for obvious ethical reasons) to how much one can engineer this in an experimental setting, but in a series of studies Tyler Stillman and his colleagues manipulated events so that some participants felt they had been socially excluded, for example, by being ostracized after a ball-tossing game.[22] Even such a modest intervention as this led to the socially excluded participants feeling that life was less meaningful. The connection between good relationships and life being meaningful goes both ways. In a further study, people who reported that life was more meaningful for them were rated by others (in a controlled setting) as being more likeable and more desirable as friends.[23] This effect of meaning in life was found to be greater than that of several other variables, including self-esteem, happiness, extraversion, and agreeableness.

A twentieth-century psychiatrist whose experience made him passionate about the need to find meaning in our lives was Viktor Frankl. Because he wrote about the importance of interpersonal relationships, his account fits here though, as we shall see, it is a particular take on such relationships.

Frankl was born in 1905 in Vienna and died there at the age of 92. He is best known for his book *Man's Search for Meaning*, which has sold some 16 million copies and powerfully describes his experiences in a series of Second World War concentration camps: Theresienstadt, Auschwitz, Kaufering, and Türkheim.[24] Reading the book, as with any account of the Holocaust, is a searing experience. If one manages to get through the horrific account of what happened in such camps, one gets to Frankl's intellectual contribution. Frankl argues that even in such awful situations (Figure 7.4), one is faced with choices. It is the choices one makes that determines who one becomes:

> The experiences of camp life show that man does have a choice of action. There were enough examples, often of a heroic nature, which proved that apathy could be overcome, irritability suppressed. Man *can* preserve a vestige of spiritual freedom, of independence of mind, even in such terrible conditions of psychic and physical stress.
>
> We who lived in concentration camps can remember the men who walked through the huts comforting others, giving away their last piece

Figure 7.4 The arrival of Hungarian Jews at Auschwitz in May 1944. Even in the most ghastly of situations, some people manage to find meaning.

of bread. They may have been few in number, but they offer sufficient proof that everything can be taken from a man but one thing: the last of the human freedoms—to choose one's attitude in any given set of circumstances, to choose one's own way.[25]

This particularity of life—that each of us has to make choices about what to do and how to treat others in the specifics of where we find ourselves—means that Frankl is suspicious of grand answers to the question 'what is the meaning of life?' As he puts it:

the meaning of life differs from man to man, from day to day and from hour to hour. What matters, therefore, is not the meaning of life in general but rather the specific meaning of a person's life at a given moment. To put the question in general terms would be comparable to the question posed to a chess champion: 'Tell me, Master, what is the best move in the world?'[26]

Frankl developed a school of psychotherapy in which humans are motivated by 'a will to meaning'. This contrasts with the earlier views of the psychoanalysts Sigmund Freud—that humans are motivated by 'a will to pleasure' (the Pleasure Principle) and Alfred Adler—that humans are motivated by 'a will to power' (a concept he adopted from Nietzsche). Frankl is therefore an existentialist.

Existentialism

Existentialism is a movement rooted in a philosophical enquiry about what it is to be human. It does not deny the importance of the natural sciences but holds that they are not sufficient for us to understand ourselves.[27] As with any important movement, characterizing it by a single feature risks caricaturing it, but if one had to provide a single term that gets to the heart of existentialism it is 'authenticity'—the notion that to find meaning each of us ultimately has to be true to ourselves, to act authentically. One thinks of Polonius' advice to his son Laertes in *Hamlet*:

> This above all: to thine own self be true,
> And it must follow, as the night the day,
> Thou canst not then be false to any man.[28]

Richard Feynman, whom we quoted in Chapter 3, followed Polonius in admonishing his hearers not to fool themselves.[29]

The roots of existentialism go back to Søren Kierkegaard (1813–55), a Danish philosopher and theologian. Kierkegaard was deeply interested in the relationship between objectivity and subjectivity—issues we touched on in Chapter 5—namely, what we can reliably know about the world that others can agree on and what we may know about the world because of who we are. Objective facts are important but what is more important for Kierkegaard is how we relate to these. Kierkegaard's writings are full of the need for us to take certain 'leaps'. One can chose whether to leap into faith as one can choose whether to leap into love.

As a philosophical movement, existentialism really took off in the mid-twentieth century in continental Europe. Fortunately for its popularization, some of its foremost advocates—Jean-Paul Sartre and Albert Camus—were fine novelists too.[30] Indeed, they were each awarded the Nobel Prize for literature. Sartre turned it down; Camus, at the age of 44, became the second youngest recipient. (Camus was also a fine goalkeeper though tuberculosis at the age of 17 put an end to any professional ambitions.)

In Sartre's *Nausea*, the central figure, Antoine Roquentin, is a loner. He has no friends and does not keep in touch with his family. The book's title refers to the boredom and lack of interest that Roquentin feels about his life. Eventually, he starts to wonder if he even exists. But the book has a positive turn. In a moment of profound realization, Roquentin appreciates that reality is existence and ultimately beyond intellect:

> I can't say that I feel relieved or happy: on the contrary, I feel crushed. Only I have achieved my aim: I know what I wanted to know; I have understood everything that has happened to me since January. The Nausea hasn't left me and I don't believe it will leave me for quite a while; but I am no longer putting up with it, it is no longer an illness or a passing fit: it is me.
>
> I was in the municipal park just now. The root of the chestnut tree plunged into the ground just underneath my bench. I no longer remembered that it was a root. Words had disappeared, and with them the meaning of things, the methods of using them, the feeble landmarks which men have traced on their surface. I was sitting, lightly bent my head bowed, alone in front of that black, knotty mass, which was utterly crude and frightened me. And then I had this revelation.
>
> It took my breath away. Never, until these last few days, had I suspected what it meant to 'exist'. I was like the others, like those who walk along the sea-shore in their spring clothes. I used to say like them: 'The sea *is* green; that white speck up there *is* a seagull', but I didn't feel that it existed, that the seagull was an 'existing seagull'; usually existence hides itself.[31]

In a 1945 lecture, Sartre came up with the phrase '*l'existence précède l'essence*' ('existence comes before essence')—a core claim of existentialism and a reversal of pre-existentialist philosophy (Figure 7.5). The phrase is a natural consequence of the argument in Martin Heidegger's magnum opus *Sein und Zeit* (*Being and Time*).[32] In this book, Heidegger uses the German word *Dasein*, which literally means 'there being', to explore what we mean by being and how it is affected by temporality.[33] Heidegger points out—an anticipation of Sartre's *l'existence précède l'essence*—that the way to get to grips with a hammer (our phraseology) is not to reflect on it but to seize it and use it. It is in hammering that we get to know a hammer. Something similar was written of the life of Jesus: 'That which was from the beginning, which we have heard, which we have seen with our eyes, which we have looked upon and touched with our hands'.[34]

Existentialism has had an influence on many psychotherapists. In *Existential Psychotherapy*, the US psychiatrist Irvin Yalom presents

Figure 7.5 A sweet chestnut tree, an important player in Sartre's *Nausea*.

four ultimate concerns of life—death, freedom, isolation, and meaninglessness—and illustrates them with reference to clinical cases. He cites the following note:

> Imagine a group of happy morons who are engaged in work. They are carrying bricks in an open field. As soon as they have stacked all the bricks at one end of the field, they proceed to transport them to the opposite end. This continues without stop and everyday of every year they are busy doing the same thing. One day one of the morons stops long enough to ask himself what he is doing. He wonders what purpose there is in carrying the bricks. And from that instant on he is not quite as content with his occupation as he had been before.
>
> I am the moron who wonders why he is carrying the bricks.[35]

This was a suicide note—the last words of someone who saw no meaning in life. Yalom argues, though, that the search for meaning is paradoxical in that one doesn't find meaning by searching for it directly—it needs to be pursued obliquely.

In the 1993 film *Groundhog Day* Phil Connors, a weatherman, finds himself stuck endlessly—and not to his liking—in Punxsutawney. Every morning it is again the same 2 February. Connors eventually decides to try to get his producer, Rita Hanson, to fall in love with him. At first, Connors tries to impress Hanson—having learnt on one 2 February what her favourite drink is, on the next 2 February he orders one too (the film is full of wonderful throw-away lines like 'How do you know there will be a tomorrow? There wasn't one yesterday'). Having learnt that she studied French poetry, he learns some French poetry.

These tactics cause Hanson to become amazed—and suspicious. Has Connors been asking her friends about her simply so that he can get her into bed with him? Eventually, Connors stops trying to bed Hanson and starts getting to know the people of Punxsutawney. This he does—very well; after all he apparently has forever. Eventually, he spends much of each 2 February being helpful, using his prior knowledge from the many previous 2 Februarys: catching one boy as he falls from a tall tree towards the pavement below; performing the Heimlich manoeuvre on a man about to choke to death; saying just the right thing to various people. Seeing one evening how much Connors is appreciated, Hanson falls in love with him. When Connors wakes the next morning, with Hanson alongside, it is 3 February.

The phrase 'groundhog day' has entered the English language and in 2006, the film was added to the United States National Film Registry as being deemed 'culturally, historically, or aesthetically significant'.[36] It has also become popular for people in many faith traditions:

> There's much to the view of Punxsutawney as purgatory: Connors goes to his own version of hell, but since he's not evil it turns out to be purgatory, from which he is released by shedding his selfishness and committing to acts of love. Meanwhile, Hindus and Buddhists see versions of reincarnation here, and Jews find great significance in the fact that Connors is saved only after he performs mitzvahs (good deeds) and is returned to earth, not heaven, to perform more.[37]

The popularity of *Groundhog Day* suggests that audiences find in it a reflection of something which they can relate to in their experience. Could that be connected to their search for meaning? And if so, could it be because for some of them the search for meaning invites exploration of the transcendent dimension of flourishing?

The transcendent dimension to meaning

Transcendent experiences often carry with them a sense of timelessness. Even people absorbed in such everyday activities as cooking or gardening can easily lose track of time. A recent article in the *Journal of Sustainable Tourism* entitled 'An Existentialist Exploration of Tourism Sustainability: Backpackers Fleeing and Finding Themselves' quotes one of the research participants:

> It's funny how the days of the week used to dominate my life and be my routine. Having this unforeseen opportunity to see the world has meant I have completely lost touch with time. It often comes as a shock to discover the day of the week and the date. Losing track of time, to just be in the moment, to feel alive, to just breathe, feel, see and hear all the splendour in this glorious world has filled my heart with so much gratitude. I am truly thankful.[38]

This feeling, of being in the moment, of time, in a sense, standing still, is close to the concept of 'flow' in positive psychology. Named by Mihály Csíkszentmihályi, flow is about 'being in the zone'. Csíkszentmihályi cites the example of a professor of physics who was also an avid rock climber: 'It is as if my memory input has been cut off. All I can remember is the last thirty seconds, and all I can think ahead is the next five minutes.'[39]

Flow is sometimes described as 'effortless attention'; 'effortless' captures something of it in that there is no way a person can induce flow simply by trying hard, but flow does require concentration. At the same time, there is a loss of self-consciousness and no fear of making a mistake. As one dancer put it:

> A strong relaxation and calmness comes over me. I have no worries of failure. What a powerful and warm feeling it is! I want to expand, to hug the world. I feel enormous power to effect something of grace and beauty.[40]

Csíkszentmihályi became interested in the concept of flow because of artists who would 'get lost in their work'. Mental states like boredom and anxiety are antipathetic to flow. Because the person is 'in the moment', time can pass very slowly. Csíkszentmihályi cites the example of ballet dancers who told him that a difficult turn that takes less than a second in 'real' time could feel as if it took minutes. Sportspeople similarly sometimes talk about 'having all the time in the world' or seeing a baseball or cricket ball (if they are batting) 'as if it was a football'.

In his *One Day in the Life of Ivan Denisovich*, Russian author Alexander Solzhenitsyn, who spent many years in Stalin's labour camps, captures both Csíkszentmihályi's idea of flow and Frankl's belief that meaning can be found anywhere. The novel follows a single day in the 3,653-day sentence of Ivan Denisovich Shukhov (the last paragraph of the novel is 'The three extra days were for leap years.'). One scene, particularly well captured in the 1970 film of the same title as the book, is to do with the intrinsic satisfaction Shukhov gets from using his building skills, despite it being slave labour and the weather being so cold that the mortar freezes within minutes:

> His thoughts and his eyes were feeling their way under the ice to the wall itself, the outer façade of the power-station, two blocks thick. At the spot he was working on, the wall had previously been laid by some mason who was either incompetent or had scamped the job. But now Shukhov tackled the wall as if it was his own handiwork. There, he saw, was a cavity that couldn't be levelled up in one row: he'd have to do it in three, adding a little more mortar each time. And here the outer wall bellied a bit – it would take two rows to straighten that.[41]

That evening, before going to sleep, Shukhov looks back on his day: 'he'd built a wall and enjoyed doing it.'[42]

Are there ever circumstances in which flourishing is not possible and lives cannot be meaningful? The philosopher John Cottingham asks us to consider a concentration camp guard who tortures the camp inmates:

> Unless the concentration camp guard proposes to turn himself into nothing more than a machine for the infliction of cruelty, he will presumably need, if only in his off-duty hours, human conversation, emotional warmth, the cultivation of friendships, family ties... Furthermore, since the sensibilities required for such human pursuits cannot be switched on and off at will, but are necessarily a matter of permanently ingrained dispositions of character, the gratification our guard is supposed to be deriving from his gruesome work will inevitably create a psychic dissonance, which will sooner or later to endanger a collapse—either a breakdown of his ability to continue as a torturer or a breakdown of his ability to live a fulfilling home life. Of course it is (unhappily) conceivable that a job that involves cruelty and bullying may produce excitements that may make it horribly attractive to certain individuals; that is not in dispute. The point is that it cannot, for the reasons just given, constitute a coherent model for a meaningful human life.[43]

This sounds convincing. And yet, even concentration camp guards may perhaps lead lives of relative goodness. Viktor Frankl, who had a certain moral authority to speak on such matters, relates how one of his camp commanders never once lifted his hand against any of the prisoners. Indeed, after liberation the camp doctor, a prisoner himself, told the other camp survivors that this commander had paid no small amount of his own money to purchase medicines for the prisoners from the nearest market town. Frankl goes on to relate:

> An interesting incident with reference to this SS commander is in regard to the attitude toward him of some of his Jewish prisoners. At the end of the war when the American troops liberated the prisoners from our camp, three young Hungarian Jews hid this commander in the Bavarian woods. Then they went to the commandant of the American Forces who were very eager to capture this SS commander and they said they would tell him where he was but only under certain conditions: the American commander must promise that absolutely no harm would come to this man. After a while, the American officer finally promised these young Jews that the SS commander when taken into captivity would be kept safe from harm.[44]

It is not the circumstances in which one finds oneself that matter in the search for meaning but who one is and what one does. Meaning is constrained by circumstances but determined by choices.

Religion

Many people find meaning in religion.[45] This doesn't mean that they don't find it in anything else but they find religion provides the ultimate source of meaning. Most religions hold that there is meaning to be found but finding it can be difficult. Jesus admonished his hearers: 'Enter through the narrow gate. For wide is the gate and broad is the road that leads to destruction, and many enter through it. But small is the gate and narrow the road that leads to life, and only a few find it.'[46]

In the Abrahamic religions (Judaism, Christianity, and Islam), meaning is primarily to be found in responding to God, the creator and sustainer of the Universe. It is difficult to generalize about any religion as all religions encompass great diversity but all three of the Abrahamic religions maintain that God is one (nuanced by the Christian understanding of the Trinity) and hold that for each of us and for worshipping communities, meaning comes from relating to God in ways that

Figure 7.6 Pentecostal worship.

are ultimately good for us and in accordance with what God wants. Thereafter, generalizations about how each religion understands meaning become difficult because of the range of alternative understandings of each religion's scriptures and other teachings.

Each of the Abrahamic religions has mystic elements. For example, in Islam, Sufism—which occurs in various Islamic denominations— seeks to enable the believer to have a direct experience of Allah. In Christianity, the Pentecostal tradition found in various denominations emphasizes the direct communication that is possible between God and believers, through such practices as speaking in tongues and prophecy (Figure 7.6). In such ways, believers feel that they have an especially close relationship with God and so are able to discern God's meaning for their lives directly.

Cottingham identifies the moral and spiritual meaning of a human life as the most important dimension of Christian belief:

> In the Christian worldview, the whole of creation is brought into being out of love; and the infinite concern of the creator is extended to every one of us. The world we inhabit it is not an impersonal world, not a world of meaningless flux where conditions arise and pass away, and where selfhood is an illusion, but a world in which it is truly said even of the sparrow, sold in the marketplace for two farthings, that 'not one is forgotten before God'.

It is a world in which we are told 'not to be afraid, for the very hairs of your head are numbered'.[47] This is the source of the exhortation and hope which, however hard it may sometimes be, the believer his urge to hold onto; it is this which makes learning to believe not merely an intellectual exercise but something which, if the promise of faith is true, will open the door to the 'joy which no man taketh from you'.[48]

For believers, religion channels their understanding of where to look for meaning. There is still the opportunity for individuality but ultimately God is seen as a reality and the source of all meaning. In the Christian tradition, in which the authors stand, this meaning is to be found in knowing God.

Knowing God

Why have so many books been written about knowing God, from John Templeton's *The God Who Would Be Known* (whose thesis is that God is revealing himself through developments in the sciences) to Jim Packer's *Knowing God* (embracing almost the whole of the Christian life of faith)?

Sir John Templeton finds evidence of God's communication through the scientific study of his universe. The opening chapter of *The God Who Would Be Known* carries the title of the book. It concludes:

Exploration of God's universe that is just beginning now becomes a new journey of spiritual discovery, a voyage into the sphere of the spirit.... The universe does indeed have meaning and purpose, and we can *read the message*. The implications for our science are staggering. For example, consider what we may learn of the power of love examined not in Freudian unbelief, but with the openness to the spiritual message that 'God is love.'[49] Again, think of hope. Is it really just a psychological projection of an unmet need, or could it be a God-given resource for the mobilisation of inner strength?[50]

This is a God who can be known because we can read the message of meaning and purpose contained in the Universe and elucidated through science. The message has content which originates from what Templeton described as the Great Revealer.

Jim Packer presents knowing God as providing a foundation, shape, and goal for lives, plus a principle of priorities and scale of values:

What were we made for? To know God. What aim should we set ourselves in life? To know God. What is the 'eternal life' that Jesus gives? Knowledge of God 'This is life eternal, that they might know thee, the only true God, and Jesus Christ whom though hast sent'.[51] What is the best thing in life, bringing more joy, delight, and contentment than anything else? Knowledge of God.[52]

Some European languages distinguish objective knowledge of a thing from relational knowledge of a person. Thus, in French *savoir* and in German *wissen* each refer to knowledge about facts and subjects. In the same languages *connaître* and *kennen* each refer to knowing a place or a person (as in the Cumberland hunting song, 'D'ye ken John Peel?'). As with any language there are finer subtleties and variations, but the distinction is a useful one.

Measuring meaning

The most important things in life can be the hardest to measure, but there have been significant advances in the metrology of meaning. This enables correlations to be drawn between meaning and independent measures of psychological flourishing, with a growing weight of evidence that people who believe that their lives have meaning are happier, enjoy greater overall well-being, and report higher life and work satisfaction and control over their lives.[53] Three aspects of meaning emerge in a definition that 'meaning in life is the set of subjective judgements people make that their lives are (a) worthwhile and significant, (b) comprehensible and make sense, and (c) marked by the embrace or pursuit of one or more highly valued, overarching purposes or missions.'[54] The last of these is closely related to purpose, which is thus seen as a distinct component of meaning.

Using a closely related classification, the *Comprehensive Measure of Meaning* instrument asks individuals to perform a self-assessment under each of seven headings:

1. Coherence: life making sense to the person living it
   A. Global: a sense of the world generally, of human life specifically, and of the ultimate meaning of life
   B. Individual: the meaning of one's own life, the capacity to understand the meaning of life events, and a philosophy that helps one understand one's identity
2. Significance: life having value, worth, and importance
   A. Subjective: a perceived subjective sense of significance of one's life as a whole, the process of living, and the kind of life one has

B. Objective: the things that one does, one's life as a whole, and one's contributions, either in what the actions are in and of themselves or to society

3. Direction: a unified understanding of what one's life should be

A. Mission: having a mission or calling, an awareness of that mission, and that mission giving one direction in life

B. Purposes: having a sense of direction or purpose, one's awareness of one's purposes, and one's more immediate goals being aligned with those purposes

C. Goals: having goals, the importance of those goals, and an awareness of those goals.[55]

Although the *Comprehensive Measure of Meaning* is designed for empirical research to assess having a sense of meaning and its causes and effects, it may also serve for individuals who wish to review the ways in which their own life has meaning.

Am I living my life in harmony with reality?

Is meaning to be found, to be chosen, or to be created? It seems to be a subtle interplay between all these three. A well-known story likens the search for meaning to a pearl trader finding the most magnificent specimen he has ever encountered in his whole professional career. With seemingly reckless disregard for spreading risk through a diverse portfolio, he mortgages everything to acquire that one pearl.[56] The pearl was real enough, but he had to make the choice.

The life coach Tom Paterson encourages his clients to recognize five domains of their lives: personal; family; faith; vocation; community. He invites them to write a specific objective for each domain. He then offers 'four helpful questions' to apply to each domain:[57]

1. *What is right about your life?*
2. *What is wrong about your life?*
3. *What is confused in your life?*
4. *What is missing from your life?*

One can use a large matrix to apply these four questions to the five domains. To each element of the matrix one could then add further questions to reflect the extent to which the individual is in control. In what ways should I seek to *discover* meaning in this element? In what ways should I seek to *choose* meaning in this element? In what ways should I seek to *create* meaning in this element?

Shakespeare identified seven different ages of humans.[58] The discoveries and choices of an individual will need to be applied afresh, or possibly even replaced, in response to the changes in their life as the years go by. In Part III we look at changes that are taking place in the world, and how scientific insight and spiritual wisdom may need to be updated, or at least applied afresh, for human flourishing in uncertain times.

Notes

1 Longfellow, H. W. (1861) Paul Revere's ride, *The Atlantic*, January. Available at https://www.theatlantic.com/magazine/archive/1861/01/paul-revere-s-ride/308349/.

2 Eliot, T. S. (2018) *The Poems*, vol. I, ed. C. Ricks and J. McCue, London: Faber & Faber, pp. 574–5.

3 Gaines, J. (2005) *Evening in the Palace of Reason*, New York, Harper Perennial, p. 237.

4 Reddaway, W. F. (1904) *Frederick the Great and the Rise of Prussia*, New York: G. P. Putnam & Sons.

5 Shakespeare, W. (1598) *Love's Labour's Lost*, Act V, Scene II, line 214.

6 Scruton, R. (2016) *The Ring of Truth: The Wagner's Ring of the Nibelung*, London: Allen Lane, p. 200.

7 Diels, H. and Kranz, W. (1985) *Die Fragmente der Vorsokratiker*, Zurich: Weidmann, DK 80B1.

8 Cottingham, J. (2002) *On the Meaning of Life*, Abingdon, UK: Routledge, p. 17.

9 Sperlich, R. (2017) Top 10 hoarders who were killed by their own hoard, Listverse, 19 June. Available at https://listverse.com/2017/06/19/top-10-hoarders-who-were-killed-by-their-own-hoard/.

10 Tolin, D. F., Frost, R. O., Steketeed, G. and Fitcha, K. E. (2008) Family burden of compulsive hoarding: Results of an internet survey, *Behaviour Research and Therapy* **46**(3), 334–44.

11 Kondo, M. (2010/2014) *The Life-Changing Magic of Tidying Up*, London: Vermilion. See also Marie Kondo's website, https://konmari.com.

12 Sooke, A. (2016) The man who destroyed all his belongings, BBC Culture, 14 July. Available at http://www.bbc.co.uk/culture/story/20160713-michael-landy-the-man-who-destroyed-all-his-belongings.

13 Ibid.

14 Anderson & Garland (n.d.) How to identify: antique mourning jewellery. Available at https://www.andersonandgarland.com/news-item/How-to-identify-antique-mourning-jewellery/.

15 National Geographic (2012) Dia de los Muertos. Available at https://www.
    nationalgeographic.org/media/dia-de-los-muertos/.
16 Parel, A. J. (2016) *Pax Gandhiana: The Political Philosophy of Mahatma Gandhi*, New
    York: Oxford University Press.
17 Mark 8: 36.
18 Bolt, R. (1980) *A Man for all Seasons*, London: Heinemann.
19 Basbanes, N. A. (1990) *A Gentle Madness: Bibliophiles, Bibliomanes and the Eternal
    Passion for Books*, New York: St Martin's Press.
20 MacKenzie, M. J. and Baumeister, R. F. (2014) Meaning in life: Nature,
    needs, and myths, in A. Batthyany and P. Russo-Netzer (Eds), *Meaning in
    Positive and Existential Psychology*, Dordrecht: Springer, pp. 25–37.
21 Lambert, N. M., Stillman, T. F., Baumeister, R. F., Fincham, F. D., Hicks,
    J. A. and Graham, S. M. (2010) Family as a salient source of meaning in
    young adulthood, *Journal of Positive Psychology* **5**(5), 367–76.
22 Stillman, T. F., Baumeister, R. F., Lambert, N. M., Crescioni, A. W., Dewall,
    C. N. and Fincham, F. D. (2009) Alone and without purpose: Life loses
    meaning following social exclusion, *Journal of Experimental Social Psychology*
    45(4), 686–94.
23 Stillman, T. F., Lambert, N. M., Fincham, F. D. and Baumeister, R. F. (2011)
    Meaning as magnetic force: Evidence that meaning in life promotes
    interpersonal appeal, *Social Psychological and Personality Science* **2**(1), 13–20.
24 Frankl, V. E. (1946/2004) *Man's Search for Meaning*, London: Rider.
25 Ibid., pp. 74–5.
26 Ibid., p. 113.
27 Crowell, S. (2015) Existentialism, *The Stanford Encyclopedia of Philosophy*, 9 March.
    Available at https://plato.stanford.edu/archives/win2017/entries/existentialism/.
28 *Hamlet*, Act I, Scene III, lines 78–80.
29 Feynman, R. (1997) *Surely You're Joking, Mr Feynman!*, New York: W. W. Norton,
    p. 343.
30 We can recommend https://existentialcomics.com. Try the 'Random' button.
31 Sartre, J.-P. (1938/1965) *Nausea*, London: Penguin Books, p. 182.
32 Heidegger, M. (1927/1962) *Being and Time*, Oxford: Blackwell.
33 Webb, B. L. J. (2021) *Science, Truth, and Meaning: From Wonder to Understanding*,
    Singapore: World Scientific.
34 1 John 1: 1.
35 Yalom, I. (1980) *Existential Psychotherapy*, New York: Basic Books, p. 419.
36 National Film Registration Board, Film registry. Library of Congress. Available
    at https://loc.gov/programs/national-film-preservation-board/film-registry.
37 Goldberg, J. (2006) A movie for all time, *National Review*, 2 February. Available at
    https://www.nationalreview.com/2006/02/movie-all-time-jonah-goldberg-2/.
38 Canavan, B. (2018) An existentialist exploration of tourism sustainability:
    backpackers fleeing and finding themselves, *Journal of Sustainable Tourism* **26**(4),
    551–66.

39 Csíkszentmihályi, M. (1975/2002) *Flow*, London: Rider, p. 58.

40 Ibid., p. 61.

41 Solzhenitsyn, A. (1962/1963) *One Day in the Life of Ivan Denisovich*, Harmondsworth, UK: Penguin Books, p. 79.

42 Ibid., p. 143.

43 Cottingham (2003) *On the Meaning of Life*, pp. 27–8.

44 Frankl (1946/2004) *Man's Search for Meaning*, p. 93.

45 Baumeister, R. F. (1991) *Meanings of Life*, New York: Guilford Press.

46 Matthew 7: 13–14.

47 Luke 12: 7.

48 Cottingham, J. (2009) *Why Believe?*, London: Continuum, p. 144. 'Joy which no man taketh from you' is from John 16: 22.

49 1 John 4: 8.

50 Templeton, J. M. and Herrmann, R. L. (1989) *The God Who Would Be Known: Revelations of the Divine in Contemporary Science*, West Conshohocken, PA: Templeton Foundation Press, pp. 15–16.

51 John 17: 3.

52 Packer, J. I. (1973) *Knowing God*, London: Hodder and Stoughton, p. 31.

53 Steger, M. F. (2012) Experiencing meaning in life: Optimal functioning at the nexus of well-being, psychopathology, and spirituality, in P. T. P. Wong (Ed.), *The Human Quest for Meaning: Theories, Research, and Applications*, 2nd edn, New York: Routledge, Chapter 8.

54 Steger, M. F. (2018) Meaning in life: A unified model, in C. R. Snyder, S. J. Lopez, L. M. Edwards, and S. C. Marques (Eds), *The Oxford Handbook of Positive Psychology*, 3rd edn, Oxford: Oxford University Press.

55 Hanson, J. A. and VanderWeele, T. J. (2021) The comprehensive measure of meaning: Psychological and philosophical foundations, in M. Lee, L. D. Kubzansky, and T. J. VanderWeele (Eds), *Measuring Well-Being: Interdisciplinary Perspectives from the Social Sciences and the Humanities*, Oxford: Oxford University Press.

56 Matthew 13: 45–6.

57 Paterson, T. (1998) *Living the Life You Were Meant to Live*, Fort Collins, CO: Paterson Center, p. 80.

58 Shakespeare, W. (c.1600) *As You Like It*, Act II, Scene VII, lines 138–65.

PART III
CHANGING CONTEXTS OF HUMAN FLOURISHING

Overview

Human flourishing is timeless. The context in which it is to be promoted is not. Therefore, our attempt to address the contemporary opportunities for human flourishing will necessarily date faster than the earlier parts of the book. Every generation will have to ask afresh how established thinking about human flourishing must be either replaced or applied in new ways to new challenges. Part III offers case studies in three different domains of change that we perceive in the world as we write. This is the part of the book which will most require updating by the reader, and application to the challenges they face. Being specific is the best route to being general: by looking at these examples we hope to illustrate how to address further challenges in the future.

As we were preparing to write this book, we were involved in a series of consultation events to identify changes in the world which would have an impact on human flourishing. We held a series of discussions in Oxford with colleagues who shared our interests and concerns, and international meetings in Rome, Paris, and London, and Kenya, Cyprus, and California. Here are some of the topics which emerged from those consultations.[1]

- The changing role of education and the university
- City planning and urban architecture
- Leisure—the forgotten path to a life well-lived
- Longevity, suffering, and vulnerability
- Building communities: going global versus going local
- Negotiating fractured identities and nurturing self-belief
- Automation and the future of work
- Active citizenship in the age of artificial intelligence
- Gene-editing

Each of these is a topic for fruitful exploration. Each offers new challenges for flourishing, and each offers new opportunities for flourishing. As we evaluated these and other changes for the three case studies in Part III, we found that three floated to the top in significance. The first is in the human sciences: the limits to predictability. The second is in human demographics: the changing patterns of religious commitment. The third is technological: of all the vast changes which are taking place we select machine learning and gene synthesis as two that ask deep questions about what it means for humans to flourish.

Life is both stochastic and non-linear. Stochasticity means that there is a certain randomness in life. As Harold MacMillan is supposed to have replied when asked what is most likely to knock governments off course, 'Events, dear boy, events.' The same applies to our decisions. I could have chosen X, but, hey, I chose Y. Non-linearity means that the consequences of even small variations in events and choices can be disproportionate to the causes. Two candidates apply for the same job. There is little to separate them in their qualifications and experience, but a persuasive voice on the selection panel speaks in favour of one of them. That person gets the job, and the other one doesn't. The difference in input is small; the difference in outcomes is total. And with a different composition of the panel the outcome could have been otherwise. Human flourishing has to be robust against such radical uncertainty.[2]

Much of contemporary economic theory is based on the assumption that humans are greedy, lazy, and selfish, and they behave rationally to achieve their ends. This model of *Homo economicus*, Economic Man (for once perhaps we should indulge lack of gender diversity), assumes that as consumers humans seek to get as much of what they want as they can for the minimum cost, and as producers they seek to make as much profit as they can for the minimum effort. This may have seemed attractive to a simple interpretation of Darwinian evolution, with a suggestion that humans who behaved like that would be selected for their survival advantage. If a criterion for a scientific theory is to be sufficiently true to be useful, then sadly *Homo economicus* is sufficiently true to be misleading. There are ways in which humans act like that; the mistake was to think that they only or mainly act like that.[3] If the social sciences are to be morally load bearing, then they need to be based on a profound understanding of what it is for humans to flourish.[4]

Religious traditions have been reflecting on what it means to be human for thousands of years. They ought to have something to contribute on the subject of human flourishing. Central to many tradi-tions, and certainly to the Christian tradition to which the authors belong, is the question 'Is this life all?' In many aspects of human flour-ishing the answer might make little difference operationally, however big a difference it may make motivationally. In other aspects, including some of the deepest ethical questions about human flourishing, it makes a huge difference. Therefore, we would expect changing pat-terns of religious commitment to be significant for promoting human flourishing, especially through the combination of scientific insight and spiritual wisdom.

Commitments to different traditions have changed significantly across the globe. Judaism has experienced diaspora for more than two millennia, with the return to the land of Israel still argued about and fought over, and antisemitism still finding unwelcome expression. Buddhists from Tibet are to be found in exile in India. Islam has long since spread beyond the Arab world, and is increasingly present in hitherto predominantly Christian countries. The notion of a post-Christian Europe is being replaced by post-European Christianity. While national cultures often still draw on the heritage of religious traditions that shaped them, the connections are becoming increasingly severed, to the extent that many citizens may scarcely be aware of them.

Alongside these demographic changes in patterns of religious commitment are intellectual changes. Where once spiritual awareness was expressed within regular patterns of institutional worship, it is becoming increasingly common for surveys to find a majority of respondents describing themselves as spiritual, but not reporting any regular pattern of attendance at religious rituals or gatherings.[5] Necessarily this kind of spirituality is not tied to traditional creedal statements or sacramental ceremonies. Nevertheless, whatever the changes in the quantity of adherents of any faith tradition, it would be premature to presume a decline in the quality of either commitment or eagerness to learn about religious possibilities. Simply by way of example, the Alpha course for enquirers which emanated from a church in central London has now seen over 24 million people in more than 100 countries around the world participate in its courses.[6] The challenge, then, is not that people have lost interest in exploring the transcendent dimension of human flourishing, but rather how to give expression to that in a material world which seems to be so well described within a scientific and technological intellectual framework.

The final chapter in Part III is about how humans flourish within the extraordinary technological changes taking place before our very eyes. We considered several candidate technologies for this chapter. Those associated with mitigating climate change are of undoubted importance. There are two kinds of reasons for caring about the environment. Most public policy about climate change is driven by the effect on humans, often on compassionate grounds for the poorest in the world who suffer earliest from effects such as drought and rising sea levels. Another family of reasons for caring might be described as transcendent, to do with valuing the material world around us for reasons quite distinct from the impact on humans. Whichever is the driving motivation, to a large extent the goals are not in dispute, only the willingness to commit to what is necessary in order to achieve them. We save climate action for the end of Chapter 11. For Chapter 10 we chose two technologies whose limits are unknown, and which are likely increasingly to demand the best of scientific insight and spiritual wisdom if they are to contribute to human flourishing in a way commensurate with their potential. We hope that readers will extend the considerations to those technologies with which they are most familiar.

Machine learning thrives on data. Anyone who uses the Internet, whether for browsing or communicating through email or social media, is offering their data to the cyberworld, which the machine learning engines of data companies harvest.[7] Anyone in Europe who visits a new website is likely to be asked whether they are willing to accept cookies. Why? Because the cookies enable the organization behind the website to harvest their browsing data. The power of machine learning is then available to learn about their preferences, and then to target influence on them with the aim of changing their preferences.[8] How will humans exercise responsibility for their preferences?

Machine *learning* is just that. Where previously the user might have programmed a computer to perform a task, now the machine is programmed to *learn* how to perform a task. Increasingly the task may consist of decision-making. Machines are surpassing humans in an increasing range of tasks, some of them involving highly complex analysis. Increasingly, questions will arise regarding which decisions are best left to machines, and which require human involvement. Where decisions require both technical and moral input, how will humans harness the strengths of the machines for the technical dimension while ensuring that the moral component remains firmly rooted in truth, purpose, and meaning?

Gene synthesis is at an earlier stage of development than machine learning, but it may come even closer to the heart of what it means to be human. Already gene editing by techniques such as CRISPR-Cas9 enables individual genes to be edited. The technique has been rightly hailed for its potential for good. Nevertheless, the announcement in 2018 by He Jiankui of the birth of twin girls whose genomes he had edited to prevent HIV infection provoked a shocked reaction, though apparently that has not inhibited further experiments by Denis Rebrikov involving gene editing of hereditary mutations associated with deafness.[9] Meanwhile, other techniques such as prime editing and base editing seem to advance apace.[10] Gene synthesis will take all this further, to the point where it will one day be possible to type in the complete genome of an organism, and synthesize the gene with sufficient fault-tolerant error correction to render the genome useful for information storage and synthetic biology.

How should we use the ability to edit and synthesize genes as the technologies become progressively available? There are practical questions of safety, which is why germline editing (for offspring yet-to-be conceived) should proceed with more caution than somatic gene editing (for that person only). There are also questions of regulation, which is about whether international standards can be agreed and how they can be enforced, in a way that will avoid unregulated clinics claiming to be able to eliminate inherited conditions by using inadequately tested and possibly harmful procedures. But even if such concerns can be satisfactorily addressed, the deeper questions will remain about how such capabilities should be used to promote human flourishing. A simple distinction between curing and enhancing proves, on closer examination, to be a difficult line to hold. We know that from other areas of health care. There is no rigorous demarcation between restorative skin surgery and cosmetic surgery. Immunization is given to healthy children not to cure MMR but to enhance their resistance to otherwise debilitating diseases. If we know that there are genetic contributions to cancer, heart disease, mental illness, and sexual preference,[11] and if one day we learn the causal links, then how should prospective parents and physicians decide on responsible genetic intervention?

The technologies of machine learning and gene synthesis are as different as could be. The details of their regulation and the ethical decisions associated with them are correspondingly specific. And yet the deeper you dig to find foundational principles, the more the questions converge. What does it mean for people to be fully human? How can humans best behave responsibly? If that is true of the disparate technological changes, it is also true of other changes which impact human flourishing. They too prompt us to ask afresh what humans are for, and how human flourishing is to be promoted through the best of scientific insight and spiritual wisdom.

Notes

1 Burbidge, D., Briggs, A. and Reiss, M. J. (2020) *Citizenship in a Networked Age: An Agenda for Rebuilding Our Civic Ideals*, Oxford: University of Oxford. Available at https://citizenshipinanetworkedage.org/wp-content/uploads/2020/04/CiNA-Report-for-Web-with-Links.pdf.

2 King, M. and Kay, J. (2020) *Radical Uncertainty: Decision-Making for an Unknowable Future*, Boston: Little, Brown.

3 Collier, P. (2019) Greed is dead: The recognition that we need to rely on each other rather than ourselves, *Times Literary Supplement*, 6 December.

4 Smith, C. (2015) *To Flourish or Destruct: A Personalist Theory of Human Goods, Motivations, Failure, and Evil*, Chicago: University of Chicago Press.

5 Ecklund, E. H., Johnson, D. R., Vaidyanathan, B., Matthews, K. R. W., Lewis, S. W. and Thomson, R. A., Jr. (2019) *Secularity and Science: What Scientists around the World Really Think about Religion*, Oxford: Oxford University Press.

6 https://alpha.org.uk/about.

7 Zuboff, S. (2019) *The Age of Surveillance Capitalism: The Fight for a Human Future at the New Frontier of Power*, London: Profile Books.

8 Russell, S. (2019) *Human Compatible: Artificial Intelligence and the Problem of Control*, London: Penguin Random House.

9 Editorial (2019) Human germline editing needs one message, *Nature* **575**, 415–16.

10 Ravindran, S. (2019) Fixing genome errors one base at a time, *Nature* **575**, 553–5.

11 Ganna, A. et al. (2019) Large-scale GWAS reveals insights into the genetic architecture of same-sex sexual behaviour, *Science* **365**, 882; see also Lambert, J. (2019) No 'gay gene': Massive study homes in on genetic basis of human sexuality, *Nature* **573**, 14–15, and Maxmen, A. (2019) Controversial 'gay gene' app provokes fears of a genetic Wild West, *Nature* **574**, 609–10.

8

Limits to Predictability

The economist Paul Samuelson was once asked by a mathematician whether he could name an idea in economics that was both universally true and not obvious. His answer was the principle of comparative advantage, an idea that was published in 1817 by the English businessman and politician David Ricardo. Ricardo portrayed an imagined scenario involving Portugal, where a certain quantity of wine could be produced by 80 workers and a certain quantity of cloth by 90 workers, and England, where the same amount of wine could be produced by 120 workers and the same amount of cloth by 100 workers.[1] Since both wine and cloth can be produced with less effort in Portugal, you might think that they should produce their own wine and their own cloth. Not so, showed Ricardo. Both Portuguese and English will benefit if the Portuguese make the wine and the English make the cloth, and the countries then trade wine for cloth so that each has what they need. What counts is not the absolute advantage (Portugal is more efficient at both) but the comparative advantage (the differences between the relative efficiencies for wine and cloth). This provides the basis for trade between countries and within countries. It also provides a basis, alongside other considerations, for sharing tasks within a household.

For a simpler example of an economic theory that makes predictions, consider Figure 8.1. This shows an idealized representation (that is, a model) of a supply curve S. As the price that consumers are prepared to pay for a certain good increases, so does the supply. The demand curves show how the quantity purchased increases as the price decreases; this is known as the law of demand. When demand is at D_1, a price P_1 equates to the quantity Q_1. Now suppose that demand increases from D_1 to D_2. The model predicts that both the price of the good and the quantity of it that is consumed increase. That is the theory. The model can be tested; it often does a pretty good job. But increasing the price of a luxury good can sometimes lead to an increase in demand, not a decline. A luxury good serves as a status symbol precisely because most

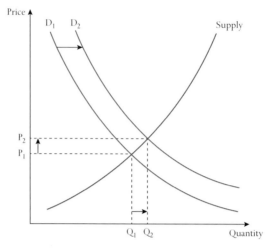

Figure 8.1 An idealized supply curve, showing how the price of a good/service and the quantity consumed are predicted to vary as demand changes.

people can't afford it. Is a £10,000 Rolex watch really that much better than a quartz crystal watch costing a hundredth of that amount?

One of the qualities of science is its ability to make predictions which can subsequently be tested. The biologist Sir Peter Medawar, Nobel Laureate for his work on tissue grafting, whose discovery of acquired immunological tolerance paved the way for organ transplants, wrote 'I expect that its embarrassing infirmity of prediction has been the most important single factor that denies the coveted designation "science" to, for example, economics.'[2] The failure of most economists to predict the global crash of 2008 led many to question the reliability of economics as a predictive science. Michael Fish presented the BBC weather forecast on 15 October 1987, during which he said, 'Earlier on today, apparently, a woman rang the BBC and said she heard there was a hurricane on the way. Well, if you're watching, don't worry, there isn't!' He was wrong. That day the South of England experienced its most extreme storm for three centuries. Severe damage occurred, and 19 people lost their lives. We did not for that reason close the Meteorological Office and abandon weather forecasts—but we can learn to make more reliable forecasts. What makes economic forecasting so difficult?

Some years ago, one of us worked with the entrepreneur and investor Hermann Hauser to explore whether option analysis would help us in our decision making about funding research. Options are widely used by hedge funds. Suppose the value of a stock today is $10. You might think that at present it is overvalued, but there is a possibility that in future it may be worth much more than $10. Rather than purchase the stock now, you may purchase at a much lower price, say $1, the right to buy the stock at an agreed price, say $9, in the future. If the stock never reaches a price at which it is worth buying, then you will have lost $1 a share. But if the price goes up to, say, $15 a share, then you can exercise your option and make a profit of $5 a share. In some circumstances that might be an attractive option.

Real options are similar, except that the option is now not a commitment on paper but a tangible asset. Suppose a piece of land in Australia has known deposits of uranium, but at present the market price of uranium would not justify the cost of mining and processing it. Nevertheless, you might decide to purchase the land, with the possibility of extracting the ore if the price in future justifies this. That would be a real option to mine the resources of that piece of land. In a real R&D option the asset is the ability to undertake research and development. A laboratory might have expertise and facilities, and perhaps also intellectual property (IP), for developing vaccines. At present the research does not seem viable, but it might foreseeably become worthwhile in the future. One could decide to keep the core skills of the team employed, and their equipment in good condition, so as to maintain the possibility of investing in a future much more substantial research effort if developments warrant it. This is a real option for research.

Hermann and Andrew investigated the possibility of using a mathematical model for real R&D option analysis similar to what is used by hedge fund traders for stock option analysis. The partial differential equation widely used for that purpose is the Black–Scholes formula, which yields, under the assumption of a frictionless market, the correct price for the option that allows one to buy and sell the underlying asset so as to hedge the ensuing risk. We found that the Black–Scholes equation could not usefully be used for our investment decisions. At its simplest, the problem was that the equation demands as an input the value of the underlying asset. For a publicly listed share that is readily available. For a commodity such as uranium the price is also listed, although it is somewhat more complicated to estimate the value of the ore in the ground. But what is the market value of a research capability? Without knowing that, it is impossible to use the Black–Scholes equation.

This illustrates a more profound point about models in general, and financial models in particular. If you allow for the uncertainties in the values of each of the inputs, you can use established methods of differential calculus to propagate those through the model to yield an uncertainty in the output. If the resulting uncertainty is greater than the uncertainty in taking a decision based on your experience and the collective wisdom of your colleagues, then the model is of no use to you. You are better off simply trusting your judgement.

A former governor of the Bank of England and a former head of the Saïd Business School have together written a book with a title which says it all, *Radical Uncertainty*.[3] Drawing on vast national and international experience of economic modelling, they show how time and again models can be misleading and that relying on them can result in unwise and sometimes harmful decisions. A common failing is that the models begin by assuming complete and perfect knowledge of the world, when in fact very few of the relevant data are known. In such situations, models may be at best useless, and at worst deeply damaging in their consequences. The refrain running through their book is that in such cases it may be much better to stand back and ask a more qualitative question: 'What is going on here?'.

In 1998 Amartya Sen won the Sveriges Riksbank Prize in Economic Sciences in Memory of Alfred Nobel, popularly known as the Economics Nobel Prize. At the time he was writing *Development as Freedom*, which was published the following year. In that book, Sen argues that development as traditionally conceptualized is too narrow, because of its focus solely on increases in gross national product or personal incomes. Sen argues that development requires a number of linked freedoms. These include certain political freedoms such as the liberty to participate in public discussion, and the removal of major restrictions on freedom such as tyranny, poverty, and neglect of public facilities. Sen recounts an experience he had as a boy:

> I was playing one afternoon—I must have been around ten or so—in the garden in our family home in the city of Dhaka, now the capital of Bangladesh, when a man came through the gate screaming pitifully and bleeding profusely; he had been knifed in the back. Those were the days of communal riots (with Hindus and Muslims killing each other), which preceded the independence and partitioning of India and Pakistan. The knifed man, called Kader Mia, was a Muslim daily laborer who had come for work in a neighboring house—for a tiny reward—and had been

knifed on the street by some communal thugs in our largely Hindu area. As I gave him water while also crying for help from adults in the house, and moments later, as he was rushed to the hospital by my father, Kader Mia went on telling us that his wife had told him not to go into a hostile area in such troubled times. But Kader Mia had to go out in search of work and a bit of earning because his family had nothing to eat. The penalty of his economic unfreedom turned out to be death, which occurred later on in the hospital.[4]

Sen describes how this devastating experience made him reflect on how economic unfreedom can breed social unfreedom. Economic predictions, and economic values, depend on an accurate understanding of what motivates humans.

For half a century or more, much economic modelling has used the concept of *Homo economicus*. This assumes that individuals always behave wholly rationally to achieve the maximum amount of what they want, whether it be desirable goods or enjoyable experiences, for the minimum cost. Humans, in this economic model, are rational, greedy, selfish, and lazy. *Homo economicus* underpins much of the economic organization of our world. How true is it? Is it sufficiently true to be useful? Or just sufficiently true to be misleading?

No doubt humans sometimes, perhaps often, do act rationally, greedily, selfishly, and lazily. We don't need footnotes to support that. We only need a modest experience of life to recognize it in others. Recognizing it in ourselves may be a bit harder, but not impossible. The mistake is to base modelling on the belief that humans *only* act like that.[5] Not everything that matters in life is amenable to free markets, because not everything can be measured in monetary value.[6] If Nobel Prizes were put up for auction, the committee would have more money to distribute and the winning bidders would presumably be pleased with what they had purchased. But the prize itself would thereby lose its significance and prestige. A community concert, in which the ticket price is deliberately kept low, would fail in its purpose if the tickets were instead sold for the highest price the market would bear. If babies were available in a free market, then both seller and buyer might be satisfied with the transaction, but somehow we know that babies should not be sold and bought like any other commodity.

In many circumstances, identity plays a decisive role in motivation and moral choices. A soldier under fire in battle is unlikely to undertake a utilitarian calculation of how to get the greatest benefit at minimum

cost. Soldiers in danger are much more likely to be motivated by loy-
alty to comrades; much of their training is designed to inculcate that
sense of identity.[7] For many years, men were substantially more likely
to smoke than women. But this cannot be explained by the sorts of
models we see in Figure 8.1. Even though men tended (and tend) to
have greater disposable incomes than women, the reason for the difference between smoking rates in men and women has far more to do
with what society expects. For many years, smoking was thought to be
manly but unladylike. That changed with the Women's Lib movement
in the 1960s. In a series of advertisements under the strapline 'You've
come a long way, baby', Virginia Slims actively equated smoking with
women's liberation (Figure 8.2).

A history of advertising shows how in times past TV adverts, at the
relevant time of day, were targeted at a woman's identity as a housewife.
The targeted identity has moved with the times but the principle is
the same. Gender identity is a complex mixture of biology, cultural
expectations, and choice, and remains powerful. Religious identity, too,

Figure 8.2 Gender differences in smoking rates illustrate the importance of
cultural norms for economics. This photograph shows two 1968 advertisements
that successfully targeted women as potential smokers.

depends in complex ways on cultural background and choice, with Article 18 of the Universal Declaration of Human Rights protecting freedom to change religion or belief. Religious identity can be very powerful in motivating behaviour that is not intended to bring primary benefit to the individual, and often involves financial and personal cost.

Corporate employment might be thought to be an area governed by the notion of *Homo economicus*. But if you believe that your staff are inherently lazy and only want to get paid as much as possible for doing as little work as possible, then you will have to put in all sorts of layers of supervision to ensure that they do what is required of them. And then *Quis custodiet ipsos custodes?* You will have to put in layers of supervision that rise even higher, culminating with the incentive package for the CEO herself. This is not restricted to industry and commerce. In a recent book with John Kay, Paul Collier recalls 'attending a dinner for the heads of eleven European civil services at which Britain's head of service proudly explained that not only were monitored incentives being adopted throughout the public sector, but he himself was on an incentive pay system, linked to targets for which the prime minister was his assessor.'[8] Collier and Kay doubt that such a senior person actually worked harder and more effectively because of his monitored incentives. They also doubt that someone who needed to be motivated by such measures would be the right person to head the civil service. One study estimated that so much of a social worker's time is taken up with the paperwork required to protect them and their superiors in the event of failure that less than 20 per cent of their time is available for actually looking after people.[9]

Evolutionary pressures and processes are now seen to be much more complex than simply the random mutation of genes and the propagation of the germline of the fittest individuals. Humans can be motivated by their commitment to the group with which they identify, whether it be their fellow soldiers on the battlefield or a social purpose to which they contribute their time. The purpose of a corporation can include increasing shareholder value, but it can extend to the benefit of all the other stakeholders, starting with its customers.[10] Economics can have an ethical basis. One of the notable features of humans is their capacity to work together to undertake complex tasks. There is growing evidence that social evolution towards increasing cooperation has been accompanied by genetic evolution to provide the physiological resources for social interaction at an ever higher level.[11] If *Homo economicus* describes a person who is greedy, lazy, and selfish, then perhaps *Homo fidelis* can be used to describe humans who are generous, energetic, and altruistic. Such a description fits with the way that humans have evolved, and it is more conducive to promoting human flourishing. It is perfectly possible to recognize the social and genetic developments that have evolved to provide us with the wherewithal for cooperation,[12] while at the same time retaining individual and collective responsibility for choosing how we deploy those capabilities.[13]

As humans cooperate, the material and the relational dimensions of human flourishing go hand in hand. Imagine that you were isolated with no one to help you and you wanted to make something as simple as a ballpoint pen. How would you go about it all by yourself? And yet with the cooperation of others, from the factory to the supply chain, you can purchase one for the cost of perhaps less than a minute of your time—if you can remember when you last paid for a ballpoint. 'Ah!' you say, 'The reason I need a pen is in order to communicate with others.' Exactly so. Now imagine that you wanted to make a smart phone, also for the purpose of communicating with others. That costs more than a few minutes of most people's time. To make the chips that go into it requires a fab line that may cost more than the GDP of a small country. In 2018 Tsinghua Unigroup started production in a fabrication plant in Nanjing that had cost $30 billion.[14] That may be small compared with the world cost of the COVID-19 pandemic, but it is large compared with most individuals' disposable income, and for sure no individual understands every detail of the plant. No other animal species can cooperate on such a complex scale. If cooperation is so essential for human flourishing, could it be that it should be taken account of in models not simply as a *means* but also as a *motivation*? That would not be easy to quantify with any useful degree of certainty.

Living with uncertainty

If you ask any school student, 'Who first observed whether two objects dropped from a given height strike the ground at almost the same time?', they may answer Galileo Galilei, and they may add that he dropped the objects in question from the top of the Leaning Tower of Pisa. Whatever the origins of that legendary story, the observation had already been made a more than a thousand years earlier by John Philoponus in Alexandria.[15] In writing 'almost the same time' he recognized that there is a limit to the accuracy of the observation. Rather better attested than dropping things from the Leaning Tower of Pisa are Galileo's experiments rolling a bronze sphere down a sloping groove. This gave a slower rate of descent than free-fall, which he could therefore time using a water clock. For each timing he weighed the amount of water 'with a very accurate balance'. Accuracy was becoming increasingly significant in experimental science. Galileo continued that he was able to determine the differences and ratios of the times 'with such accuracy that although the operation was repeated many, many times there was no appreciable discrepancy in the result.'[16]

This healthy obsession with accuracy was taken to a new level by members of the *experimentall philosophicall clubbe* which met in the rooms of John Wilkins, Warden of Wadham College, Oxford, from 1649 onwards each Thursday at two o'clock for what we would now call the weekly seminar. The club included scientists (although they were not called that then) familiar to this day: Edmund Halley (whose eponymous comet helps us to date the Battle of Hastings to 1066); Christopher Wren (now better known for his architecture); Robert Hooke (after whom Hooke's Law is named, and who at the time was best known for his *Micrographia*), and Robert Boyle.

In 1660 the Honourable Robert Boyle, younger son of the Earl of Cork, published his first scientific work, *New Experiments Physico-Mechanicall, Touching the Spring of the Air and Its Effects*. He described how he used a **J** tube with the short end sealed and the longer end open to the atmosphere. By pouring mercury into the open end, he could compress the air in the closed end and compare the length of the region containing air with the difference in height of the mercury in the two sections of the tube. In his tabulation of the results he gives the length of the air column in the shorter leg, the height of the column in the longer leg, and then the effective height after adding 29 1/8 inches of mercury to allow for atmospheric pressure (note the precision of 1/8 inch). In the final column of the table he writes 'What the pressure should be according to the *Hypothesis*, that supposes the pressures and expansions to be in reciprocal proportion.'[17] That is what we now call Boyle's Law. There is no fixing of results. The figures in the last two columns are close, but not identical. It may be that one of the sources of error was fluctuations in the diameter of the **J** tube in the region where the gas was confined. Boyle appraises whether this denotes agreement or disagreement, concluding that the variations could be attributed to 'such want of exactness as in such nice experiments is scarce avoidable.' Evaluation of errors is foundational to the apprenticeship of every experimental scientist.

On 4 July 2012 the scientific world was electrified by the announcement of evidence for the Higgs boson in experiments at the Large Hadron Collider at CERN, Geneva. The CERN team was professional and effective at communicating the uncertainty associated with the discovery. Even those who were not familiar with standard deviations learned that this was a five-sigma result (sigma, σ, being the Greek symbol for standard deviation), upgraded from previous tantalizing two- to three-sigma results. They learned that 5 sigma meant that there was a probability of about one in three million that this is a false result. The CERN team, and plenty of other science communicators, were careful to explain that this did not mean that there was a 99.999967 per cent probability that the Higgs boson exists, but rather that there was a 0.0000033 per cent probability of getting the result they did even if it does not exist. Even those for whom this was overstretching their grasp of statistics nevertheless hopefully got the point that science is not a matter of certainty, but rather of measured uncertainty. This was an important achievement in the public communication of how science is done.

The appreciation of uncertainty is central to science. Science is forever developing. We recognize breakthroughs in science whenever previous ways of thinking are superseded. Occasionally, previously accepted ideas are found to be, well, just wrong. But that is relatively rare in science. More often models which have been found to yield reliable results within certain limitations need to be modified for situations outside the constraints within which they were developed. Newton's laws of motion serve us well for large objects moving much slower than the speed of light. For very small objects and objects moving very fast, and for timekeeping for the precision required by GPS satellites, we need the further developments of twentieth-century physics. A privilege and a responsibility of a professor in a research university in science, as in other academic disciplines, is to identify areas of uncertainty in our knowledge that are likely to be amenable to significant progress within the scope of a doctoral project.

This uncertainty in science can readily be misunderstood. Just because within the history and philosophy of science breakthroughs have occurred that showed previous understanding to be either wrong or incomplete does not mean that we cannot be confident about anything, or, to be more precise, sufficiently confident to use it as the basis for decisions. We can be sufficiently confident of the phase diagrams of aluminium alloys to entrust our lives to them every time we fly in an aeroplane. We can be sufficiently confident of germ theory to know how to reduce the likelihood of infectious disease transmission.

Human flourishing requires a nuanced appreciation of uncertainty. If we are not sure of anything then we cannot make informed decisions, we cannot implement wise choices, and we cannot take responsibility for the ensuing consequences. But if we are too sure of everything, then we shall be living with unwarranted confidence in assumptions that will be unable to bear the load that we place on them. There is a saying that has been attributed to Woody Allen: 'If you want to make God laugh, tell him about your plans.' If the world had forgotten the wisdom of that, then the COVID-19 pandemic provided a forceful reminder. Any promotion of human flourishing must take account of what is outside our control and outside our capacity to know, and of the radical uncertainty of the world in which we exercise our responsibility as humans. How does this work out in different areas of scientific endeavour?

The physical sciences

Everyone knows that in space light travels in straight lines. In 1915 Albert Einstein produced his theory of general relativity in which if that is so then space must be curved. One of general relativity's predictions was that paths of light would appear to be bent by the presence of objects. In everyday life, the amount of bending is predicted to be minimal but when we are talking about the effect of our Sun (a truly massive object) on the light from more distant stars, the apparent bending should be appreciable. Isaac Newton had wondered whether objects might bend rays of light. However, he wasn't able to make any quantitative predictions about the extent of such bending and when later physicists did, within a Newtonian framework, the prediction turned out to be half the extent of bending compared to that predicted by Einstein's theory of general relativity.[18]

The published version of Einstein's theory of general relativity reached the UK in 1916, in the middle of the First World War—not the best time for scientific expeditions. But two British astronomers, Arthur Eddington (Director of the Cambridge Observatory) and Frank Watson Dyson (the Astronomer Royal), realized that when such expeditions again became possible, the total solar eclipse of 1919 would provide the opportunity to test one of the predictions of Einstein's theory.

During this eclipse, it was known (as we mentioned in Chapter 1, we can predict eclipses to great accuracy) that the Sun would appear through a telescope to be close to the cluster of bright stars known as the Hyades. A photograph would show the apparent position of the Hyades relative to other stars not close to the Sun. This apparent position of the Hades could be compared to its apparent position relative to these other stars when the Sun is nowhere near them in the sky. Any difference in the two apparent positions should be due to the curvature of space in the vicinity of our Sun, as observed through the light travelling from the Hyades to us as observers on the Earth. Measuring the extent of this difference would allow a test of Einstein's theory.

So far so good. However, there were a number of problems. For a start, it was known that the total eclipse would not be visible from the UK or even nearby. The path of totality (as it is called) was known (calculated—again, a confident prediction) to pass from Brazil eastwards to West Africa. Expeditions to either of these regions were impossible during the war. Fortunately, the Armistice on 11 November 1918 gave just enough time to make the arrangements before the eclipse on 29 May 1919. Another potential problem—as anyone who has ever tried to watch an eclipse knows—is that success relies on good weather. In the event, the crucial measurements were taken. The 100th anniversary of the eclipse led to a number of carefully researched books that tell the tale of what happened and have helped refine the simple story that 'Eddington proved Einstein's theory of general relativity to be correct.'[19]

To maximize the chances of success, two expeditions were launched: Eddington headed off to the Island of Príncipe, off the West Coast of Africa, and Andrew Crommelin, from the Royal Greenwich Observatory, went to Brazil. Each of the expeditions had a number of logistical and other problems—bad weather, a strike, blurred photographic plates— but enough of the crucial measurements were made and the results were found to be more in line with the predictions of Einstein's theory than of Newton's. The results were announced at a special joint meeting of the Royal Society and Royal Astronomical Society in London and made headline news around the world (Figure 8.3). The *New York Times*' comfortingly declared: 'Stars Not Where They Seemed or Were Calculated to be, but Nobody Need Worry'.[20]

REVOLUTION IN SCIENCE.

NEW THEORY OF THE UNIVERSE.

NEWTONIAN IDEAS OVERTHROWN.

Yesterday afternoon in the rooms of the Royal Society, at a joint session of the Royal and Astronomical Societies, the results obtained by British observers of the total solar eclipse of May 29 were discussed.

The greatest possible interest had been aroused in scientific circles by the hope that rival theories of a fundamental physical problem would be put to the test, and there was a very large attendance of astronomers and physicists. It was generally accepted that the observations were decisive in the verifying of the prediction of the famous physicist, Einstein, stated by the President of the Royal Society as being the most remarkable scientific event since the discovery of the predicted existence of the planet Neptune. But there was difference of opinion as to whether science had to face merely a new and unexplained fact, or to reckon with a theory that would completely revolutionize the accepted fundamentals of physics.

Figure 8.3 The triumphalist announcement on the front page of *The Times* of 7 November 1919 of the verification of Einstein's theory of general relativity by British astronomers.

None of the plates from either expedition seem to have survived so it
has not been possible to check the measurements that were made. This
is unfortunate, because there has been a long-running rumble of
suspicion about the measurements and Eddington has been accused of
'cooking the books'.[21] This is probably unfair, though it is the case that
Eddington left out of the published results some that didn't agree with
his main conclusion. There were reasons for this—scientists often
explain that 'that experiment didn't work and produced anomalous
findings'—but even at the 1919 special joint meeting, there were some
scientists who questioned the findings and their interpretation.

The history of science is not always quite as straightforward as is
commonly presented. But in this case science made predictions which
could be tested and verified with high precision. Subsequent measurements of other eclipses have confirmed Einstein's predictions—though
agreement was not finally reached until the 1960s when observations were
made at radio frequencies rather than in the visible spectrum. In time,
Einstein's theory of general relativity may be supplanted just as Newton's
theory of gravitation eventually was, not least because we still do not
know how to reconcile quantum theory with general relativity. Nevertheless, the reliability of GPS bears testimony to how science can offer predictability at its most accurate. But even within the physical sciences there
are important areas which have not yielded to determinism.

Chaos theory

The artist Alexander Calder is renowned for his mobiles (the name
given to them by the artist Marcel Duchamp) (Figure 8.4). Most people
who have seen them are entranced by Calder's mobiles, even if they
may find it difficult to explain why they are considered great art. Some
of Calder's mobiles have an interesting property. If you pull or push
them from their equilibrium position the path they follow before they
eventually settle down is never the same, however much you try to
repeat exactly how you pull or push them and even if you do it indoors
where there is no wind.

This isn't a property just of Calder's mobiles; it is a widespread phenomenon. If you look up 'double pendulum' on the Internet, you can
find plenty of videos showing such pendula in motion.[22] The movement of the pendulum very quickly becomes unpredictable. In the language of physics, a double pendulum shows strong sensitivity to initial
conditions. Indeed, it is so sensitive to initial conditions that, to all

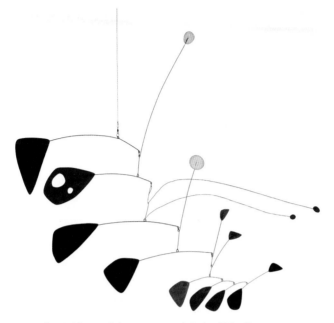

Figure 8.4 The Calder mobile *Antennae with Red and Blue Dots*.

intents and purposes, it is not practical to set the pendulum going iden-
tically on two occasions. As a result, its behaviour is unpredictable; its
motion is chaotic. Chaos is so fundamental that it has its own branch of
mathematics—chaos theory. One of us even published a widely cited
paper on the smallest chaotic mechanical system ever made.[23]

A popular name for this is 'the butterfly effect'—the idea that
precisely where a hurricane strikes might have been affected by a
butterfly beating its wings elsewhere in the world a few weeks earlier.
A 2004 science fiction film called *The Butterfly Effect* explores this idea.
Unfortunately, most critics don't rate it very highly. We prefer a Ray
Bradbury short story called 'A Sound of Thunder' first published in
1952. The year is 2055, time travel is feasible, and there is company that
regularly runs trips which enable rich hunters to go back into the past
and shoot animals that it is known would have died only a few minutes
later, minimizing the effect on subsequent events. You can probably see
where this is heading. Without giving too much away, one trip goes
slightly wrong and on returning to 2055 there are subtle changes—
people behave a bit differently, English is spelt and sounds differently,

and there is a different President of the United States. Looking at his boots, the hunter finds a crushed butterfly.

Bradbury's short story is often credited with popularizing the 'the butterfly effect', though the phrase was first used by Edward Norton Lorentz, a US mathematician and meteorologist. In 1961 Lorentz was using an early digital computer to simulate weather patterns—he would enter all sorts of data for things like wind speed and temperature and see what weather was predicted. One day he wanted to examine a sequence at greater length. Since the computer was slow, he took a short cut. Instead of starting the whole run from the beginning, he used numbers from the printout from halfway through the previous run. This should not have affected the outcome but, to his surprise, the end result was utterly different.[24]

Lorentz's first thought was that the computer had malfunctioned; it hadn't. Then he thought of another explanation—the right one. In the computer's memory, numbers were stored to six decimal places, e.g. 0.506127, but on the printout, to save space, numbers were written to only three decimal places—0.506 in this case. It was the printout data that Lorentz entered. At the time, everyone presumed (it wasn't even a stated assumption) that a difference of one part in a thousand would be insignificant. But it turned out that this small difference was crucial. In 1987 Michael Fish inadvertently showed how difficult it could be to produce a reliable forecast for the next 24 hours, and the ability to forecast weather with useful accuracy very rapidly becomes more demanding the further ahead you try to do it. By using ever more measurement stations and ever larger supercomputers the extent of detailed forecasts has improved over the past forty years by about one day every ten years. If that rate of improvement is sustained, we can expect about an extra week ahead by the end of the twenty-first century.

Quantum measurement

We cannot predict when a radioactive atom will decay. That is why the clicks of a Geiger counter follow a random pattern. We may know the half-life—the time taken for half the atoms in a collection of radioactive atoms to disintegrate—to great accuracy. This is equivalent to saying that by the end of a half-life there is a 50 per cent chance that the atom will have decayed—and there is a 25 per cent chance that after two half-lives it still won't have decayed, a 12.5 per cent (one in eight) chance that after three half-lives it won't have decayed, and so on. Radioactive decay is probabilistic.

Most things in physics and chemistry are probabilistic. Following the discovery of quantum mechanics at the beginning of the twentieth century, we now believe that indeterminacy at the very small scale is a central phenomenon of all of matter. It had long been presumed that improved measurements would (in principle) give us an ever more accurate picture of where any atom in the Universe was and, if in motion relative to the observer, in what direction it was going, how fast, and whether it was accelerating. Now we know, for example, that the more accurately we know the speed and direction of motion of an object, the less precisely we can know where it is. Heisenberg's uncertainty principle provides the mathematical formulation of this. It usually doesn't matter for large objects, but at the atomic level it can be crucial.

Erwin Schrödinger and Albert Einstein didn't like some of the more radical interpretations of quantum theory and Einstein pointed out that according to one of the most well-known of these radical interpretations (the Copenhagen interpretation), two apparently unconnected physical states could still yield correlated measurements. Schrödinger called this *Verschränkung*.[25] It is how you would describe your arms when they are folded in front of your chest. The usual English translation is *entanglement*. He devised a *Gedankenexperiment* (thought experiment) which has become known as Schrödinger's cat. He conceived the thought experiment during the 1930s when he lived in Oxford on the same road where Andrew now lives. Schrödinger's cat was called Milton, after whom Julia Golding named one of the time-travelling companions in the children's book series based on *The Penultimate Curiosity*; in the series Milton can be indecisive and he sometimes jumps to unpredictable conclusions.[26] Schrödinger never did his experiment on Milton and you should not attempt it at home or anywhere else.

In his thought experiment, the cat is in a (sound-proofed and opaque) steel chamber along with the following which, Schrödinger pointed out, need to be out of the cat's reach: a Geiger counter (to detect radioactivity), a hammer, and a glass flask that contains the poison hydrocyanic acid (Figure 8.5). The Geiger counter has a 50 per cent chance each hour of being triggered by the decay of a radioactive substance. If the Geiger counter is triggered, it causes the hammer to fall which smashes the glass flask, resulting in the death of the cat. So the question is, by the end of an hour, is the cat dead or alive? According to what became known as the Copenhagen interpretation, based on

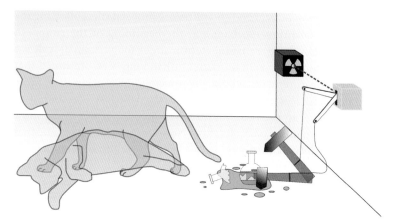

Figure 8.5 The Schrödinger cat experiment. At any one time is the cat alive, dead, or both?

ideas developed there by Niels Bohr and Werner Heisenberg between about 1925 and 1927, the cat would not be in a definite state until after the box had been opened—until then it would be simultaneously alive and dead. Einstein and Schrödinger argued that this made no sense. If one opened the chamber and found a dead cat, a forensic pathologist would be able to tell one approximately when the cat died. At any moment in time, as far as Einstein and Schrödinger were concerned, the cat is either dead or alive.

Quantum theory has been tested to perhaps greater precision than any other theory devised by the human mind, about one part in a billion. It may therefore seem amazing that quantum experiments can readily be devised in which the uncertainty in correctly predicting the outcome is no better than tossing a coin. This points to another kind of uncertainty associated with quantum theory, which is what it actually means. In Chapter 5 we opened Part II by describing how Niels Bohr and Albert Einstein were unable to reach agreement about their rival interpretations of quantum theory. In 1957 another physicist, Hugh Everett, introduced what became known as the many-worlds interpretation.[27] In this understanding of reality, at the point at which the chamber is opened, one world continues in which the cat is alive and another world continues in which the cat is dead. The two worlds are 'decoherent'—that is, no communication is possible between them. Viewers of Star Trek, Doctor Who, and various sci-fi films should have no problem recognizing such independent time lines. Advocates of the many-worlds interpretation argue that it is the only way to account simultaneously for the contents of quantum mechanics and the appearance of the world. Sceptics wryly observe that while it may be economical in explanations, it is expensive in universes.

After nearly a century of using quantum theory with spectacular success, there is still no consensus among its practitioners as to what it means.[28] Quantum theory thereby offers a profound lesson in living with uncertainty about foundational questions. For example, one could reasonably assert that measurement is indispensable to scientific inquiry. But in quantum science there is no agreement about what a measurement is. How, one might ask, can a scientist of integrity continue to measure quantum systems when they cannot even give a widely accepted account of what happens when they make a measurement? The answer, of course, is that quantum theory is too robust, and technologies that depend on it are too important, to abandon the endeavour pending a philosophical breakthrough.

Just so with uncertainty in other dimensions of human flourishing. In the relational dimension, it is not necessarily (perhaps seldom) those who are best able to articulate the neuropsychology of marriage who make the best spouses. In the transcendent dimension, most religions value the place of prayer. There is no agreed account of what happens when we pray, and the uncertainty is not diminished by using a technical term such as 'special divine action'. How, one might ask, can the husband continue to love his wife when he cannot give a satisfactory account of his affective motivations, and how can the believer continue to pray when she cannot give an empirically testable account of how God responds? The lesson from science (if it needs to be learned from science) is that there can be things that are too robust and too important to put off just because of uncertainties about underpinning mechanisms and explanations. That is not a reason for ceasing to inquire, but it is an argument for not allowing uncertainties to prevent us from getting on with life and thereby promoting human flourishing.[29]

The biological sciences

By and large, biological phenomena are less predictable than physical ones. It may be that quantum effects are important—it has been argued that they play a central role in photosynthesis, smell, vision, enzyme activity, and a number of other biological processes—but even if they aren't, there are a number of reasons why biology is far from deterministic.

Biological phenomena tend to be probabilistic rather than exact.[30] This is particularly the case with more complex phenomena. Consider, for example, the chances that a pair of songbirds will successfully rear young one year (Figure 8.6). There are many, many factors that affect breeding success in songbirds. For a start, it depends on the fitness of each parent bird; first-time breeders are generally less successful than birds that have bred before. Where the nest is built is important; for the American robins shown in Figure 8.6, the most successful nests are hard for predators to locate.[31] The time of year when the eggs are laid is important; too early or too late and there won't be as much food. There

Figure 8.6 Many factors affect the breeding success of American robin (*Turdus migratorius*) pairs, some of which are down to chance.

are less predictable factors like the weather, and so on. Much of prediction in biology is probabilistic; it is about tendencies, likelihoods.

Biological effects can be so complex that it can be hard to work out what is causing what. It might be that at least part of what is going on is deterministic but it's difficult to figure out. A classic example is the story of wolves in the Yellowstone National Park.[32] Humans have, by and large, viewed wolves as undesirable—they are seen as dangerous and don't feature well in fiction (think *Little Red Riding Hood*, *The Three Little Pigs*, *Peter and the Wolf*, *The Wolves of Willoughby Chase*, and Maugrim in *The Lion, the Witch and the Wardrobe*).[33] Back in the 1930s, wolves were exterminated from Yellowstone National Park. As a result, elk (red deer) increased in numbers to the point where they were having an adverse effect on tree regeneration.

In 1995, amid considerable controversy, wolves were reintroduced. Almost the first effect was that the elk started to forage in different places—places where it was harder for the wolves to catch them. As a result, tree and shrub regeneration happened very quickly in the places where the elk were now much rarer. This regeneration led to more songbirds and migratory birds and to more beavers. At the time of the wolf reintroduction, there was a single colony of beavers in the park; by 2019 there were nine. The trees reduced soil erosion and the beavers changed the course of the rivers, creating a greater range of habitats and benefitting otters, fish, and other species. It had been the case that the main cause of mortality of the elk was deep snow. This led to a boom-and-bust cycle with some winters where there was little elk mortality and others where there was masses. With the advent of the wolves, mortality is more evenly spread out both within and across years. This has benefitted scavengers like ravens, bald-headed eagles, and bears—and bears also benefitted from more berries on the regenerating shrubs.

Some of these results were anticipated—the biologists were pretty sure that the wolves would reduce deer numbers and that this would lead to more tree regeneration as it's well-known that wolves eat deer and that too many deer prevents trees from regenerating. But many of the consequences were unanticipated. Even though we now know why the introduction of wolves led to more beavers and ravens, this wasn't expected at the time.

Biologists disagree about how predictable the outcome of evolution is. The palaeontologist Stephen Jay Gould dreamt of rewinding the tape of life and playing it again (taking a metaphor from how speech and music were recorded in those days).[34] Gould reckoned that in such a scenario the outcome would be utterly different second time around. Not so, cried Simon Conway Morris, whose data from the Burgess Shale Gould had used. If the tape of life were to be played again, Conway Morris argued, the result would be rather similar. *Convergence* describes the observation that life seems to have homed in (converged) on similar structures and processes through routes that are unconnected by evolutionary history. Conway Morris draws on a vast amount of fossil and contemporary evidence for convergence.[35] Rather like wolves causing tree regeneration, once you see it, it is impossible to unsee it.

Convergence arises because of the laws governing the world in which life arises. Take, for example, vision. The process of seeing involves light, and since Maxwell formulated his electromagnetic equations and showed that they led to waves with the same speed as light, we know that they describe how light behaves. Actually, we probably knew enough about the refraction and focussing of light even before Maxwell, from Fermat and his predecessors like Newton and Snell. To make a seeing organ with resolution limited only by diffraction you need lenses and receptors. You can either have one lens and an array of receptors or an array of lenses, each with its own receptor. That's basically it. You have then exhausted the possibilities allowed by physics. There are advantages in having at least two such organs—either for stereoscopic vision if you are a predator or to see on both sides if you might be the victim of predators. There is plenty of evidence that both kinds of seeing organs (eyes) have arisen several times independently on different branches of Darwin's tree of life. We might even say that the very processes of evolution itself give a major additional constraint, what Andreas Wagner called the arrival of the fittest.[36] There is uncertainty in the random processes of evolution, but the outcomes of the processes are constrained.[37]

Evolutionary psychology

There are a myriad of disciplines that try to understand why humans behave as they do. What distinguishes evolutionary psychology from other approaches is the presumption that humans, including our behaviours, are the products of evolution, principally natural selection.[38] Evolutionary psychologists generally assume that behaviours that are widespread can be understood as being adaptive.

Let's start off by considering something that isn't too controversial—fear of snakes. A snake appears in the Garden of Eden and their shape means that psychoanalysts have long presumed that dreaming about snakes doesn't indicate a budding interest in herpetology but is something to do with male sexual desire. Be that as it may, many people have a fear of snakes and to an evolutionary psychologist the most likely explanation is simply that many of the world's snakes are poisonous so that an innate tendency to avoid them is likely to be evolutionarily advantageous.

The psychologist Arne Öhman has spent over 40 years working on people's fear of snakes. He points out that constrictor snakes were predators of our small ancestors some 100 million years ago, even before snakes evolved venom (probably about 60 million years ago).[39] To this day, close to 100,000 people die annually from snake bites. The worst that happens with an excessive fear of snakes is that one wastes time and gets stressed, whereas failure to fear snakes may be fatal.

Arguments of evolutionary psychologists about gender roles are more controversial. Early writings that drew on evolutionary psychology tended to presume that the standard gendered patterns of behaviour that we see in the West—stereotypically, men go out to work and women remain at home and bring up the children—could now be understood as steeped in our evolutionary biology. Unsurprisingly, such ideas were widely criticized as being sexist. It is worth looking at some data, both in traditional societies and in contemporary ones.

We can't go back in time and see what men and women spent their time doing thousands of years ago—and the archaeological record doesn't help—but we can look at today's hunter-gatherer societies. In such societies, food is obtained by hunting animals and gathering plants and other edible materials—such as mushrooms and oysters. Until agriculture was invented, about 10,000 years ago, that was how all humans obtained their food. Hunter-gatherer societies tend to be more egalitarian than agricultural ones. Marshall Sahlins, who was later to criticize E. O. Wilson's sociobiology, was the first person to argue that hunter-gatherers worked fewer hours and enjoyed more leisure than typical members of industrial society.[40] Subsequent analyses have shown that people in agricultural and industrial societies work around 35 per cent more hours than do people in hunter-gatherer societies.[41]

Hunting is more often undertaken by men and gathering by women, though there are hunter-gatherer societies where women hunt the same prey as men, sometimes hunting alongside men.[42] In some hunter-gatherer societies women provide most of the calories; in others men do.[43] Mark Dyble and colleagues studied two hunter-gatherer societies: the Agta (in the Philippines) and the Mbendjele BaYaka (in the Congo).[44] In both societies they found that men and women played equal roles in determining who got to be in the camp. However, this was not the case in a comparison society—the agricultural Paranan. Here, men played a greater role than women in determining camp composition. The conclusions of Dyble and his colleagues are in line with what is found elsewhere. In hunter-gatherer societies women are not infrequently as influential and powerful as men.[45]

In modern, non-traditional societies, we find considerable variation in the extent to which roles are defined by gender. The data in Figure 8.7 show, for 17 countries and the OECD average, the percentage of adult men and women in paid employment. The countries with the greatest sex-specific difference are at the left; those with the least at the right. We draw two conclusions from these data. The first is that in every country men are more likely to be in paid employment than women—though the difference is small in some countries. The second is that there are big differences between countries.

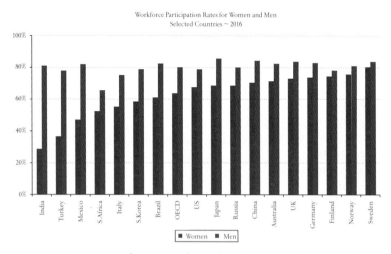

Figure 8.7 Percentage of women and men (aged 15–64 years) in the workforce in selected countries and across the OECD as a whole (data mostly from 2016).

The variation between countries is likely due to culture—for example, India and Turkey as compared to the Nordic countries. Does an evolutionary explanation help in understanding why men are (apparently) always more likely (on average) to be in paid employment than women? In principle, this too might simply be the result of cultural factors. However, it may be that in this case in particular, and for evolutionary psychology in general, there are biological roots to human behaviours which are overlain by culture.[46] It is women, not men, who become pregnant and it is women, not men, who can breastfeed. There is thus a predisposition for gendered patterns of childcare and paid employment to exist. The principle of comparative advantage is not always followed; in many societies women end up with most of the childcare *and* are in paid employment, often on substantially lower hourly rates of pay than men. Gender differences in employment rates and childcare arrangements have narrowed in recent decades—as a result of government policies and individual and couple choices. It matters how people decide what to value, and it matters what they decide.

Sociology

In Isaac Asimov's *Foundation* sci-fi series of the 1950s, Hari Seldon, a mathematics professor, develops *psychohistory*—which combines sociology, statistics, and history to make generalized predictions about the future. Psychohistory is based on two premises. The first is that the population whose behaviour is being modelled is sufficiently large; while one can't accurately predict the behaviour of one person, one can predict the average behaviour of a large number of people (though the behaviour of different demographic groups in some countries during the coronavirus pandemic has shown just how difficult that can be). The second premise is that the population should not know what these predictions are because that would cause changes in behaviour leading to unwanted feedback, thus invalidating what is technically known as the principle of stationarity.[47] Seldon helpfully leaves behind hologram messages to enable humanity, after his death, to steer itself through various crises.

The polyonymous Frenchman Isidore Marie Auguste François Xavier Comte (1798–1857) would have approved of Hari Seldon. Comte is generally seen as the founder of sociology—the study of society, including how people interact, whether at the individual, local, national, or global level. He didn't see history as a separate intellectual discipline—more as the outworking of sociology. Comte became convinced that sociology could be as much a science as physics, chemistry, or biology. However, he felt that sociology could be more than that. He saw it as the final science and its task therefore includes coordinating the development of the whole of knowledge.[48] Comte therefore felt that a sociologist needed a good understanding of the natural sciences, especially biology—the first science to deal with organized beings.

Something of the optimism of Comte's approach occasionally resurfaces. Some of its most enthusiastic supporters come from the biological sciences. In 1975 the biologist E. O. Wilson, whose specialisms were in well-established biological fields like entomology (the study of insects—ants are Wilson's especial love) and island biogeography, produced a massive book, *Sociobiology*,[49] which helped usher in a new field of biology. Here he argued that the same basic laws of genetics and evolution can explain sociality in all organisms—whether we are talking about insects, birds, or mammals, including humans. But Wilson went further. The subtitle of his book was *The New Synthesis*; Wilson saw sociobiology as the route by which biology could take over other disciplines. As he put it: 'Sociology and the other social sciences, as well as the humanities, are the last branches of biology waiting to be included in the Modern Synthesis.'[50]

Sociobiology proved to be deeply controversial. Some of Wilson's own academic colleagues at Harvard, including Stephen Jay Gould and Richard Lewontin, joined the 'Sociobiology Study Group', an organization explicitly set up to counter sociobiological explanations of human behaviour. The group published an open letter titled *Against 'Sociobiology'*, where their key objection to this new field was its deterministic outlook:

> These theories have resulted in a deterministic view of human societies and human action...we are presented with yet another defense of the status quo as an inevitable consequence of 'human nature.'...What Wilson's book illustrates to us is the enormous difficulty in separating out not only the effects of environment (e.g., cultural transmission) but also the personal and social class prejudice of the researcher. Wilson joins the long parade of biological determinists whose work has served to buttress the institutions of their society by exonerating them from responsibility for social problems.[51]

Both Wilson's book and the rapidly developing field of sociobiology were criticized in other respects too—it was widely held that they put too much weight on the importance of genes in determining how humans behave. A rapid, book-length response to Wilson's 1975 book was provided by the anthropologist Marshall Sahlins in his 1976 *The Use and Abuse of Biology*. As one might expect from an anthropologist, Sahlins' criticism was primarily to do with the way in which Wilson's theorizing led to a homogenization of human culture:

> biology, while it as an absolutely necessary condition for culture, is equally and absolutely insufficient: it is completely unable to specify the cultural properties of human behavior or their variations from one human group to another.[52]

Wilson expanded on his arguments in his 1978 book *On Human Nature*,[53] maintaining that everything from aggression and altruism to sexual behaviour and religion could best be understood through the lens of evolutionary biology. Although the book won the 1979 Pulitzer Prize for General Nonfiction, this did not pacify his critics. Forty years later, the debate has still not gone away but few sociologists are confident about the potential for their discipline to be a science like biology, even less like chemistry or physics. In Azimov's *Foundation* series, Seldon's predictions eventually fail when a man known only as 'The Mule' is born. The Mule is a one-off, a mutant with the ability to manipulate people's emotions. Seldon hadn't been able to predict this. The Mule uses his unique ability to create total devotion in his subjects and start a war, contrary to Seldon's predictions. In real life as in fiction, matters are just too complicated to allow accurate predictions to be made for all but the most mundane of human behaviours.

Uncertainty in life and in death

Life is stochastic. It is also non-linear. Stochastic means randomly determined, so that an outcome may be determined statistically but not predicted precisely. Every day we encounter circumstances that could easily have been otherwise. Human flourishing is helped if we can live life in a way that is robust to such fluctuations. Anyone learning to ski is trained to distribute their weight so that when they encounter variations in the snow, such as bumps or ice, they are able to absorb the change without taking a tumble. If they fail, for example by having their weight too far back or insufficiently on the lower ski, it will not be long before they come a cropper. Human flourishing requires skiing through life in a way that is robust against fluctuations in the surface of the piste.

Non-linear means that the effect may not be directly proportional to the cause. Much of elementary physics is concerned with effects that are linear, such as force and acceleration, or voltage and current. But not all. The power supply of your laptop converts the alternating current from your electricity to supply to the direct current needed for the battery and the transistors through the non-linear behaviour of a Schottky barrier. This is more than simply saying that small causes can have big effects, though that is undoubtedly the case. The big effects can be out of all proportion to the causes.

Anyone who has sat on an appointment panel knows how much care and thought can go into making a good decision, especially if there are two or more candidates with comparable strengths against the selection criteria. A decision has to be made, and the panel does its best. And yet time and again a tiny difference could have changed the outcome. One of the unsuccessful candidates could have presented their CV better, or performed more winsomely in the interview, or the chosen candidate might have withdrawn, or might not have accepted the job, or if the panel composition had been different another candidate might have been favoured, or if the timing had been different other candidates might have applied or these candidates might not have applied...if, if, if. That is the stochastic element. Now comes the non-linear element. Suppose that it is very hard to choose between the candidates; to make it quantitative, imagine that on a single scale there is less than 1 per cent difference between them. But one candidate gets the job—a '1' if you like—and the others don't—a '0'. That is a 100 per cent, not a less than 1 per cent, difference in outcomes.

Each of us (speaking for the authors—but we suspect it is true for you too) can identify many points in our lives when a small change would have led to a life very different from the one which we now lead. It could have been a circumstance, or it could have been a choice; either way had it been otherwise we would not be living life as it is for us now. We might have chosen a different subject to study; we might have a different job, wife, home, children...We might, we might, we might. These are not small differences, they are huge, but they depend in non-linear ways on small details which could easily have been otherwise.

As any of us looks round at our circle of acquaintances and family, we may reckon that there are some things about us that most of them know, and perhaps other things that not so many of them know. Both of us have been at meetings where in introducing ourselves around the table we were invited to say one thing about ourselves that few people knew—we shan't tell you what we disclosed! But there is one thing about us that everyone knows: we shall die. Knowing that death, alongside taxes, is one of the most certain things in life, it is surprising how little account of it most people take.

The radical uncertainty comes not in the fact of our death but in the circumstance and the timing of it. There can be various shocks in our observations of our contemporaries as we mature; we may remember the first time a contemporary married, the first time a contemporary divorced, perhaps the first time a contemporary died. One of us (Andrew) remembers the first time a contemporary had an adult child die, the first time two different contemporaries had a child die on the same day (it was a Monday, and the two deaths were entirely unrelated), and the first time a contemporary lost a child through suicide. As we pause to reflect, almost with a kind of spiritual anger, how wrong it is for a parent to bury their own child, so we also reflect that none of us knows when our own death will occur or how it will happen. Human flourishing requires living in the knowledge of that uncertainty.

What will happen then (Figure 8.8)? Will it be as Augustus Toplady expressed it, at a time when there was less of a taboo about life after death than there is now:

> While I draw this fleeting breath,
> When mine eyes shall close in death,
> When I soar to worlds unknown,
> See Thee on Thy judgment throne, … ?[54]

Or will it be as Bertrand Russell expressed it, 'I believe that when I die I shall rot, and nothing of my ego will survive. I am not young and I love life. But I should scorn to shiver with terror at the thought of annihilation.'[55] We cannot but admire that courage, although in the same essay Russell also wrote that physical science is approaching the stage when it will be complete, and therefore uninteresting. Even scientists who agree with Russell about death would, we think, demur at that. But one can contrast the attitude of Russell with that of Henry

Figure 8.8 Sixth-century Byzantine mosaic of Christ separating the sheep from the goats at St Apollinare Nuovo, Ravenna.

Venn, whose daughter was convinced that his joy at the prospect of leaving the world kept him alive an extra fortnight![56]

Beliefs about what happens after death vary both in content, from Russell to Venn, and in confidence, from the utterly certain to those who find it hard to be sure. The evidence is sparse, some would say non-existent (Chapter 4), and many years ago it was observed that even if someone should return from the dead there would be those who would not be convinced. Given the importance of the topic for human flourishing, it is remarkable how much less time and thought most people devote to it than where they will spend their summer holiday.

Values

As we argued above, human behaviour is affected by our evolutionary past but not determined by it. However, before we leave the realm of actuality—what is—and consider the normative question of what ought to be, we need to look more closely at current understandings of what is 'natural' for human behaviour in terms of our values. How we live depends on our values, and our consistency in following them.

A widely read book on AI stated that we must impart human values to the machines,[57] as though there were one thing called human values. Well, yours and mine of course, provided you agree with mine. Where we disagree . . . I am not so sure about yours.

Experimental economics allows human behaviours and the values behind our behaviours to be explored in situations that can include moral challenges. For example, a version of the dictator game has two players who do not know each other and know that they will play just a single game. One player (the dictator) is given a sum of money, say $10. This person then gets to choose how much of the money to keep and how much to give to the other person. The prediction from classical economics—that people behave rationally in their own interests—is that the dictator should keep all the money for themselves. But most people don't. Typically, they give about 15 per cent to 35 per cent to the other person.[58]

In a related game called ultimatum, one of the two players has the job of deciding how the money should be split and the other player can either accept the offer or reject it. If they accept the offer, the person who makes the offer gets to keep the rest (so, for example, if there is $10 available and the first person offers $3 and this offer is accepted, the person making the offer gets to keep the remaining $7). But if the person rejects the offer, not only do they get nothing, the person who makes the offer gets nothing too. The prediction from classical economics— again, that people behave rationally in their own interests—is that whatever you are offered you should accept it. After all, something is always better than nothing. But research shows that in this game, if less than about 20 per cent of what is available is offered, the offer is routinely rejected.[59]

It seems that people have a deep sense of fairness which may have evolutionary origins in the benefits of cooperation.[60] At its simplest, this is why we often see predators cooperating in their hunting. Depending on what prey are available, lionesses sometimes hunt on their own and sometimes in groups. When in groups, each lioness in the group may benefit as groups can go for larger prey and are more successful in such hunts than are individuals hunting on their own. There are limits, which may be why we never see lionesses hunting in groups of a dozen; any benefit in group hunting success would be more than outweighed by the large number of mouths to feed. There is growing evidence that as humans grow up they develop what we might term 'pro-social behaviours'—tendencies to cooperate.[61] But, as every parent can observe, they also have tendencies to be selfish—few people in the dictator game split the money 50:50. In a healthy society, selfishness is likely to take more of a back seat and levels of trust and respect will be higher. But it is all too easy for societies to fail. Corruption, high levels of inequality, and oppression can all cause people to behave in their narrower interests and to lose trust in others. The take home message is that it is up to us—not as solitary individuals but as individuals working together in communities—to put in place the circumstances that allow societies to promote those values that enable flourishing.

Arriving at shared values does not happen by itself. It has to be worked at. At the micro level you can hear the process at the bus stop or in the coffee break: 'I was upset by what he did'; 'I admire her for behaving that way'; 'they shouldn't do that'. It begins in the family, where through countless case studies we share our values and hopefully refine them. A healthy national discourse works towards shared values and a shared commitment to their implementation. One of the sadnesses of the UK Brexit referendum in 2016 was that, although it led to a democratic decision of a kind, there was agreement neither about what it meant nor about how to implement it, and little convergence of values as a result. As we write this chapter there is even disagreement about whether the government should be bound by an international treaty which it signed less than a year ago. Politics that promotes polarization is unlikely to lead to shared values.

Where do values come from? How do they emerge? How do they evolve? These questions are fundamental, and it requires expertise from a wide range of disciplines to address them, from philosophy and the natural sciences to economics and other social sciences. It seems to be widespread that humans, in a vast range of cultures with a diverse set of histories, agree that there is such a thing as right and wrong. And it is remarkable how many different religious traditions, as well as many other ethical traditions, have some equivalent of the Golden Rule, which can be colloquially stated as 'Do as you would be done by' or in its negative form 'Don't do to others what you would not want them to do to you.'[62] In a study of 414 societies spanning the past 10,000 years from thirty regions around the world, using fifty-one measures of social complexity, and four measures of supernatural enforcement of morality, it was found that a belief in moralizing gods followed, rather than preceded, the rise of social complexity.[63] This does not prove that there is an objective moral reality, but it does seem to show that as societies become more complex, so they become more aware of the need for enforceable moral values. And when humans become aware of a need, it is always worth looking to see what could meet it.

Within the Judaeo-Christian tradition two values underpin all the others. They are conventionally described as commandments, but since most people outside the military are not in the habit of being commanded to do things, let's call them values. The first value is to love the Lord your God with all your heart, with all your life, and with all your mind; the second value is to love your neighbour as yourself.[64] Even those who reject the first value may be willing to accept the second. What does it mean to implement it? Long ago a lawyer, wanting to win a point, posed the question 'but who is my neighbour?' The question was answered at the time by a story which became so well known that the phrase 'Good Samaritan' has entered the language.

Even when we are confident of our values, there will be all sorts of uncertainty about how to put them into practice, and about how to live peacefully alongside those who only partially share them. Nowhere is there greater need for a judicious working together of scientific insight and spiritual wisdom. Advances in science can help us to be well informed about foreseeable consequences of actions, and to make full use of technological resources for achieving our goals. But scientific insight by itself does not provide values. It certainly cannot supply all the answers in a world where values are changing and uncertain. How are we to draw on spiritual wisdom when patterns of religious commitment are also changing? To that we turn next.

Notes

1 Ricardo, D. (1817) *On the Principles of Political Economy and Taxation*, London: John Murray, pp. 158–60.

2 Medawar, P. B. (1984) *The Limits of Science*, Oxford: Oxford University Press, p. 4.

3 Kay, J. and King, M. (2020) *Radical Uncertainty: Decision-Making for an Unknowable Future*, London: Bridge Street Press.

4 Sen, A. (1999) *Development as Freedom*, Oxford: Oxford University Press, p. 8.

5 Collier, P. (2019) Greed is dead: The recognition that we need to rely on each other rather than ourselves, *Times Literary Supplement*, 6 December. Available at https://www.the-tls.co.uk/articles/greed-is-dead/.

6 Sandel, M. (2012) *What Money Can't Buy: The Moral Limits of Markets*, New York: Farrar, Straus and Giroux.

7 Akerlof, G. A. and Kranton, R. E. (2010) *Identity Economics*, Princeton, NJ: Princeton University Press.

8 Collier, P. and Kay, J. (2020) *Greed is Dead: Politics after Individualism*, London: Allan Lane, p. 23.

9 Cottam, H. (2018) *Radical Help*, London: Virago.

10 Business Roundtable (2021) Statement on the purpose of a corporation. Available at https://system.businessroundtable.org/app/uploads/sites/5/2021/02/BRT-Statement-on-the-Purpose-of-a-Corporation-Feburary-2021-compressed.pdf

11 Christakis, N. (2019) *Blueprint: The Evolutionary Origins of a Good Society*, New York: Little, Brown Spark.

12 Henrich, J. (2017) *The Secret of Our Success: How Culture Is Driving Human Evolution, Domesticating Our Species, and Making Us Smarter*, Princeton, NJ: Princeton University Press.

13 Christakis (2019) *Blueprint*.

14 Yoshida, J. (2017) Is China ready for a memory chip fab?, *EET Asia*. Available at https://www.eetasia.com/is-china-ready-for-a-memory-chip-fab/.

15 Philoponus, J. (517/1991) *Corollaries on Place and Void*, trans. D. J. Furley and C. Wildberg, London: Duckworth, p. 59.

16 Galilei, G. (1638/1914) *Dialogues Concerning Two New Sciences*, New York: Macmillan, p. 213.

17 Boyle, R. (1662) *A Defence of the Doctrine Touching the Spring and Weight of the Air, Propos'd by Mr. R. Boyle in his New Physico-Mechanical Experiments; Against the Objections of Francisco Linus. Wherewith the Objector's Funicular Hypothesis is also examin'd.* London: Printed by F. G. for Thomas Robinson Bookseller in Oxon. Reproduced in *The Works of the Honourable Robert Boyle*, London: Printed for J. and F. Rivington (1772), p. 159.

18 Coles, P. (2019) Einstein, Eddington and the 1919 eclipse, *Nature* **568**, 306–7.

19 Cowen, R. (2019) *Gravity's Century: From Einstein's Eclipse to Images of Black Holes*, Cambridge, MA: Harvard University Press; Kennefick, D. (2019) *No Shadow of a Doubt: The 1919 Eclipse That Confirmed Einstein's Theory of Relativity*, Princeton, NJ: Princeton University Press; Stanley, M. (2019) *Einstein's War: How Relativity Triumphed amid the Vicious Nationalism of World War I*, New York: Dutton.

20 *The New York Times*, 10 November 1919.

21 See the discussions in Collins, H. M. and Pinch, T. (1998) *The Golem: What You Should Know about Science*, Cambridge, UK: Cambridge University Press; and Cole, P. (2011) Einstein, Eddington and the 1919 eclipse, arXiv:astro-ph/0102462. Available at https://cds.cern.ch/record/489163/files/0102462.pdf.

22 There's an example at https://www.youtube.com/watch?v=U39RMUzCjiU&app=desktop.

23 Burnham, N. A., Kulik, A. J., Gremaud, G. and Briggs, G. A. D. (1995) Nanosubharmonics: The dynamics of small nonlinear contacts, *Physical Review Letters* **74**, 5092.

24 Gleick, J. (1987) *Chaos: Making a New Science*, London: Cardinal.

25 Schrödinger, E. (1935) Die gegenwärtige Situation in der Quantenmechanik, *Naturwissenschaften* **23**, 807–12, 823–8, 844–9.

26 Golding, J., Briggs, A. and Wagner, R. (2018) *Cave Discovery: When Did We Start Asking Questions?*; *Greek Adventure: Who were the First Scientists?*; *Rocky Road to Galileo: What is Our Place in the Solar System?*; *Hunt with Newton: What are the Secrets of the Universe?*; (2019) *Victorian Voyages: Where Did We Come From?*; *Modern Flights: Where Next?*, Oxford: Lion Hudson.

27 Everett, H. (1957) Relative state formulation of quantum mechanics, *Reviews of Modern Physics* **29**(3), 454–62.

28 Schlosshauer, M., Kofler, J. and Zeilinger, A (2013) A snapshot of foundational attitudes toward quantum mechanics, *Studies in History and Philosophy of Science Part B—Studies in History and Philosophy of Modern Physics* **44**, 222–30.

29 Briggs, G. A. D. (2014) The search for evidence-based reality, in *The Science and Religion Dialogue: Past and Future*, ed M. Welker, Frankfurt am Main: Peter Lang, pp. 201–15.

30 Watts, F. and Reiss, M. J. (2017) Holistic biology: what it is and why it matters, *Zygon* **52**, 419–41.

31 Yen, C.-F., Klaas, E. E. and Kam, Y.-C. (1996) Variation in nesting success of the American robin, *Turdus migratorius*, *Zoological Studies* **35**(3), 220–26.

32 Farquhar, B. (2019) Wolf reintroduction changes ecosystem in Yellowstone, *Yellowstone National Park Trips*. Available at https://www.yellowstonepark.com/things-to-do/wolf-reintroduction-changes-ecosystem.

33 Reiss, M. J. and Tunnicliffe, S. D. (2011) Dioramas as depictions of reality and opportunities for learning in biology, *Curator* **54**, 447–59.

34 Gould, S. J. (1989) *Wonderful Life: The Burgesss Shale and the Nature of History*, New York: W. W. Norton.

35 Conway Morris, S. (2015) *The Runes of Evolution: How the Universe Became Self-Aware*, West Conshohocken, PA: Templeton Press.

36 Wagner, A (2015) *Arrival of the Fittest*, New York: Penguin Random House.

37 Monod, J. (1970) *Le Hasard et la Nécessité: Essai sur la philosophie naturelle de la biologie moderne*, Paris: Éditions du Seuil; English translation (1971) *Chance and Necessity: Essay on the Natural Philosophy of Modern Biology*, New York: Alfred A. Knopf.

38 Downes, S. M. (2018) Evolutionary psychology, *The Stanford Encyclopedia of Philosophy*, 5 September. Available at https://plato.stanford.edu/archives/spr2020/entries/evolutionary-psychology/.

39 Öhman, A. (2009) Of snakes and faces: An evolutionary perspective on the psychology of fear, *Scandinavian Journal of Psychology* **50**(6), 543–52. For a more recent, book-length treatment of the subject, see Nobuyuki, E. (2019) *The Fear of Snakes: Evolutionary and Psychobiological Perspectives on Our Innate Fear*, Singapore: Springer Nature.

40 Sahlins, M. (1968) Notes on the original affluent society, in R. B. Lee and I. DeVore (Eds), *Man the Hunter*, New York: Aldine, pp. 85–9.

41 Sackett, R. (1996) *Time, Energy, and the Indolent Savage: A Quantitative Cross-cultural Test of the Primitive Affluence Hypothesis*, PhD thesis, University of California, Los Angles.

42 Biesele, M. and Barclay, S. (2001) Ju/'Hoan women's tracking knowledge and its contribution to their husbands' hunting success, *African Study Monographs*, Supplement **26**, 67–84.

43 Hunn, E. S. (1981) On the relative contribution of men and women to subsistence among hunter-gatherers of the Columbia plateau: A comparison with the *Ethnographic Atlas* summaries, *Journal of Ethnobiology* **1**(1), 124–34.

44 Dyble, M. Salali, G. D. Chaudhary, N., Page, A., Smith, D., Thompson, J. et al. (2015) Sex equality can explain the unique social structure of hunter-gatherer bands, *Science* **348**(6236), 796–8.

45 Endicott, K. (1999) Gender relations in hunter-gatherer societies, in R. B. Lee and R. Daly (Eds), *The Cambridge Encyclopedia of Hunters and Gatherers*, Cambridge, UK: Cambridge University Press, pp. 411–18.

46 For a fuller discussion of the different predictions of evolutionary psychology and biosocial models of gender, see Zhu, N. and Chang, L. (2019) Evolved but not fixed: A life history account of gender roles and gender inequality, *Frontiers in Psychology* **10**, 1709.

47 Kay, J. and King, M. (2020) *Radical Uncertainty: Decision-Making for an Unknowable Future*, London: Bridge Street Press, pp. 35–6, 341.

48 Bourdeau, M. (2018) Auguste Comte, *The Stanford Encyclopedia of Philosophy*, 8 May. Available at https://plato.stanford.edu/archives/sum2018/entries/comte/.

49 Wilson, E. O. (1975) *Sociobiology: The New Synthesis*, Cambridge, MA: Belknap Press of Harvard University Press.

50 Ibid., p. 4.

51 Allen, E., Beckwith, B., Beckwith, J., Chorover, S., Culver, D. et al. (1975) Against 'Sociobiology', *The New York Review of Books*, November 13. Available at https://www.nybooks.com/articles/1975/11/13/against-sociobiology/.

52 Sahlins, M. (1976) *The Use and Abuse of Biology: An Anthropological Critique of Sociobiology*, Ann Arbor: University of Michigan Press.

53 Wilson, E. O. (1978) *On Human Nature*, Cambridge, MA: Harvard University Press.

54 Toplady, A. (1775) in *The Gospel Magazine*. Available at https://www.gospelmagazine.org.uk/publications.html. Reprinted with changes to the words used today.

55 Russell, B. (1925) *What I Believe*, London: Kegan Paul, Trench Trubner. Reprinted by Routledge (2004), p. 7.

56 Foss, D. B. (1988) The Reverend Henry Venn. *Huddersfield Local History Society Newsletter*, 7, p. 3. Available at https://huddersfield-local-history-society.us-east-1.linodeobjects.com/HLHS%20Newsletter%2007%20(1988).pdf.

57 Shanahan, M. (2015) *The Technological Singularity*, Cambridge, MA: MIT Press.

58 Cummins, D. D. (2016) The surprising role of fairness in economic decision-making, *Scientific American*, MIND Guest Blog. Available at https://blogs.scientificamerican.com/mind-guest-blog/the-surprising-role-of-fairness-in-economic-decision-making/.
59 Ibid.
60 Christakis, N. (2019) *Blueprint: The Evolutionary Origins of a Good Society*, New York: Little, Brown Spark.
61 See ibid., as well as Ridley, M. (1996) *The Origins of Virtue*, New York: Viking; and Bloom, P. (2013) *Just Babies: The Origins of Good and Evil*, London: Bodley Head.
62 Templeton, J. M. (1999) *Agape Love: A Tradition Found in Eight World Religions*, West Conshohocken, PA: Templeton Press.
63 Whitehouse, H. et al. (2019) Complex societies precede moralizing gods throughout world history, *Nature* **568**, 226–9.
64 Matthew 22: 37–40, The Great Commandment; Wright, N. T. (Trans.) (2011) *New Testament for Everyone*, London: SPCK.

9

Religion and Human Flourishing

In 1851 two newlyweds passed through Dover to take a boat to Calais. The bride was Frances, daughter of Sir William and Lady Wightman. The man who was now her husband was the poet Matthew Arnold. Sixteen years later he published 'Dover Beach' in his collection *New Poems*. After an opening resonant with thoughts of honeymoon, the poem describes 'the grating roar of pebbles which the waves suck back' and continues:

> The Sea of Faith
> Was once, too, at the full, and round earth's shore
> Lay like the folds of a bright girdle furled.
> But now I only hear
> Its melancholy, long, withdrawing roar,
> Retreating, to the breath
> Of the night-wind, down the vast edges drear
> And naked shingles of the world.

Arnold's words express in verse the perception that religious faith has long been in decline. A rigorous analysis of changing patterns of churchgoing in the UK tells a story which at first glance confirms that observation.[1] The percentage of the population going to church on a Sunday was 11.7 per cent in 1979, 9.9 per cent in 1989, 7.5 per cent in 1998, and 6.3 per cent in 2005.[2] Similar trends are apparent in different measures in the 2011 UK Census and successive British Social Attitude Surveys (1983–2018).[3]

Christianity is not the only religion with adherents in the UK. Islam has grown strongly in numbers in recent years. In 2001 there were 1.55 million Muslims in England and Wales; in 2011 this number had increased to 2.71 million,[4] and an estimated 3.4 million in 2017.[5] Not everyone who identifies as Muslim regularly attends a mosque, but the attendance at mosques in the UK has been growing.

Even for those who identify as Christians, while church attendance and other quantitative measurements of church membership (such the number of baptisms and Christian weddings and funerals) have decreased, it is more difficult to ascertain trends in the other aspects of religiosity. There are data on people's self-reported measures of religious faith, but there are difficulties here in responses to questions of surveys in different contexts. A question about belief—such as 'Do you believe in God?'—may be understood differently now than in previous generations, just as scientists or philosophers might interpret the question in different ways from a broader sample of society. Data on changes in religious experiences are even patchier and less robust.

'Religiosity' is a sociological term for 'religiousness'. In everyday parlance, we may talk of someone as being 'religious', which is what the term 'religiosity' seeks to capture. It aggregates three components: religious belief, religious experience and religious practice. One person might have a strong personal faith—but only rarely go to a church, synagogue, mosque, temple or other place of collective worship. Another person might be regular in attending collective worship—but give it little thought and find it difficult to talk of their own beliefs.

Descriptions of religious belief and religious practice have existed down the centuries, but analysis of religious experience was largely a reaction to the increasing eighteenth- and nineteenth-century growth of Enlightenment rationalism in the West. The US philosopher and psychologist William James (1842–1910), often described as the 'father of American psychology', identified four characteristics of a religious or mystical experience:

- It is transient, so that the person soon returns to normal existence;
- It is ineffable, so that the experience cannot adequately be rendered into normal language;
- It is noetic, in the sense that it deepens the person's understanding of spiritual matters; and
- It is passive, in the sense that the experience cannot be obtained as a result of human intention, though certain practices, such as meditation, make it more likely.[6]

In Chapter 4 we related how various people have described their experiences of the transcendent, some of which had a religious context. Many people with a strong personal faith and who regularly engage in religious practices do not have such experiences—for them it may be more about trust and commitment and lifestyle, and their experience of the divine may be mediated through the friendship and support of a community. This is one of the advantages of the term 'religiosity'—one doesn't have to tick every box to be described, and to see oneself, as religious. There is now a whole area of research about the manifestation of religiosity. Some researchers have suggested that there are more than the three components above, of belief, experience, and practice.[7]

We have referred to religion throughout this book, as appropriate to each chapter, because it is impossible to account for the whole human story without it. In the simultaneously secular and religious cultures in which many of us live and work, many of our practices are undertaken without any explicit reference to God. And yet if one seeks to make sense of what is going on in the world without taking account of religion, the resulting understanding will be seriously deficient. In this chapter we address the issue of religion head on and ask, 'What is the contribution of religion to human flourishing?' Few questions about humanity are more contentious. For many religious believers, their religious faith is core to their being—a key component of how they understand themselves and the world as a whole; it is therefore central to their flourishing. For some atheists, religion is an evil that does great harm and holds humanity back from fulfilling its potential; they may advocate that religion needs to be debunked so that we can live free of its shackles.

Our age is simultaneously secular and religious. Few business meetings in Europe begin with a prayer, although the UK Parliament and the US Senate do, meetings of Templeton philanthropies do, and such a practice can seem quite normal in parts of Africa. Much of life in much of the world carries on with little or no explicit reference to God. But imagine how different humanity's astonishing achievements (not to mention wars and squabbles) would have been without religious motivation, from cave paintings to cathedrals, from music to literature, from schools to hospitals, and even the progress of science.

Patterns of religious commitment are changing, as is the willingness to accept the heritage of religious commitment. The Universal Declaration of Human Rights is widely regarded as a secular document, but there were strong Christian voices in its formulation. The preamble to the Treaty of Lisbon acknowledges the cultural, religious, and humanist inheritance of Europe.[8] The whole human rights movement is motivated by the value of the dignity of the human individual. Where did that come from?[9]

Changing patterns of religiosity

The Pew Research Center has a well-deserved reputation for gathering rigorous data about religiosity. In various countries people over the age of 18 were asked 'How important is religion in your life?', from a choice of 'very important', 'somewhat important', 'not too important', and 'not at all important'. Figure 9.1 shows the percentage who answered 'very important'.

The inter-country differences in Figure 9.1 are large. It is possible that in the countries with the highest figures, people may be reluctant to state that religion is not very important to them, even when faced with an interviewer who claims to be collecting anonymous data. Conversely, it probably takes considerable courage in countries like China to answer that one's religion is very important to one's life. Although the figures may overstate international differences, across the major religious traditions it seems that religion is less important for people in the Nordic countries or China and Japan than in many countries in Africa and the Americas.

Countries where population sizes are growing fastest tend to have higher levels of religiosity. Figure 9.2 shows the size in 2015 of the world's major religious groups with projections by the Pew Research Center for 2060. The figure for 'Unaffiliated' is predicted to decrease from 16 to 12.5 per cent. This is only a prediction, but a projection of the world's population who are religious increasing from 84 per cent in 2015 to 87.5 per cent in 2060 hardly fits with a narrative of religion in terminal decline. The projection suggests that it is the mainstream religions of Islam and, to a lesser extent, Christianity that are expected to increase most. Rather like the loss in linguistic diversity over time, as languages go extinct, there is projected to be a decrease in religious diversity—though this might be counteracted by increases in within-religion diversity if new denominations and sects arise and grow.

The trope that we are living in an age of increasing secularization has been around a long time and has been addressed in various ways by theologians and sociologists. In a play on words from Arnold's poem, the Sea of Faith Network was launched in 1984 following a television series and book of the same name by the radical theologian Don Cupitt.[10] Members of the network are attracted by the notion of a non-realist understanding of religion, by which is meant that we have to make our own religious knowledge; God is real for those who believe in

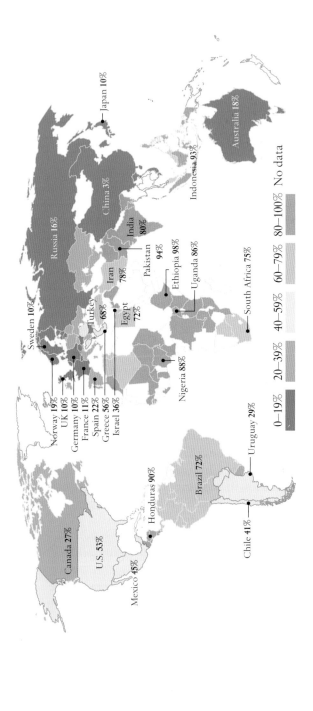

Figure 9.1 The percentage of people over the age of 18 in various countries who say that religion is 'very important' to them. Data from Pew Research Center surveys 2008–17.

	Projected 2015 population	% of world population in 2015	Projected 2060 population	% of world population in 2060	Population growth 2015–2060
Christians	2,276,250,000	31.2	3,054,460,000	31.8	778,210,000
Muslims	1,752,620,000	24.1	2,987,390,000	31.1	1,234,770,000
Unaffiliated	1,165,020,000	16.0	1,202,300,000	12.5	37,280,000
Hindus	1,099,110,000	15.1	1,392,900,000	14.5	293,790,000
Buddhists	499,380,000	6.9	461,980,000	4.8	−37,400,000
Folk religions	418,280,000	5.7	440,950,000	4.6	22,670,000
Other religions	59,710,000	0.8	59,410,000	0.6	−290,000
Jews	14,270,000	0.2	16,370,000	0.2	2,100,000
World	7,284,640,000	100.0	9,615,760,000	100.0	2,331,120,000

Figure 9.2 Projection by the Pew Research Center for changes in the sizes of the world's religions between 2015 and 2060.

him, so that to believe in the Creator is to resolve to treat life as a pure gift.[11]

The courageous and deeply intellectual German theologian Dietrich Bonhoeffer, who was hanged by the Nazis only days before the liberation of the concentration camp in which he was incarcerated, argued that a 'nonreligious' interpretation of the Gospel is needed in order to speak of God in a secular fashion. He was followed in this by Harvey Cox, who, while not a 'death of God' theologian, sought to find effective ways of communicating what religion is all about within secularization.[12] Charles Taylor, in *A Secular Age*, gives a helpful summary of what he elsewhere called the *fragilization* of belief:

> the change I want to define and trace is one which takes us from a society in which it was virtually impossible not to believe in God, to one in which faith, even for the staunchest believer, is one human possibility among others. I may find it inconceivable that I would abandon my faith, but there are others, including possibly some very close to me, whose way of living I cannot in all honesty just dismiss as depraved, or blind, or unworthy, who have no faith (at least, not in God, or the transcendent). Belief in God is no longer axiomatic. There are alternatives.[13]

The sociologist Linda Woodhead suggests that the rise in militant atheism—illustrated by Richard Dawkins' *The God Delusion*[14] and Christopher Hitchens' *God Is Not Great*[15]—was perhaps a reaction to increasing talk of desecularization.[16] It has long been observed that people often turn to religion in times of crisis. A popular saying in World War 1 was that there are no atheists in the trenches, echoing something attributed to Plato along the lines that there are few so obstinate in their atheism that a pressing danger will not reduce them to an acknowledgment of the divine power. The onset of the COVID-19 pandemic was associated, in the UK, with 5 per cent of adults saying they had started to pray during the lockdown but didn't pray before, thereby increasing the figure of adults who said that they prayed regularly (at least once a month) from 21 to 26 per cent.[17] In a culture where atheists sometimes seem to have the loudest voices, these figures suggest that disbelief in God is no longer axiomatic. There are indeed alternatives.

Does religion make the world a better place?

Christopher Hitchins is one of the group of so-called new atheists who became known as the four horsemen of the non-apocalypse, following the filming of a two-hour unmoderated discussion between them at Hitchins' home in 2007. In a book published the same year, Hitchins related the following story:

> A week before the events of September 11, 2001, I was on a panel with Dennis Prager, who is one of America's better-known religious broadcasters. He challenged me in public to answer what he called a 'straight yes/no question,' and I happily agreed. Very well, he said. I was to imagine myself in a strange city as the evening was coming on. Toward me I was to imagine that I saw a large group of men approaching. Now—would I feel safer, or less safe, if I was to learn that they were just coming from a prayer meeting? As the reader will see, this is not a question to which a yes/no answer can be given. But I was able to answer it as if it were not hypothetical. 'Just to stay within the letter "B," I have actually had that experience in Belfast, Beirut, Bombay, Belgrade, Bethlehem, and Baghdad. In each case I can say absolutely, and can give my reasons, why I would feel immediately threatened if I thought that the group of men approaching me in the dusk were coming from religious observance.'[18]

From a more scholarly perspective, Jim Jones, who is both a professor of religious studies and a clinical psychologist, addressed the intersection of religion and violence by investigating what psychological science might tell us about the phenomenon of religious terrorism.[19] He studied the Japanese cult Aum Shinrikyo, which is based in Buddhism and whose members were convicted of releasing the toxic sarin gas into the Tokyo underground in 1994 and 1995; the extreme apocalyptic elements, based in Christianity, of the religious right in the USA; and other religious movements. Jones Identified certain themes common to religiously sponsored terrorism: teachings that evoke shame and humiliation; the demand for submission to an overly idealized but humiliating institution, text, leader, or deity; a patriarchal religious milieu; an impatience with ambiguity and an inability to tolerate ambivalence that lead to a splitting of the world into all-good versus all-evil camps; a drive for total purification and perfection; narcissistic rage; doctrines that link violence and purification; and the repression of sexuality. Turning this around is not easy and reforming those convicted of religious terrorism is not straightforward.[20]

There is no doubt a large catalogue of atrocities have been carried out in the name of religion. In Oxford there is a stone cross set into Broad Street commemorating two occasions when bishops were burned alive. In colonialism and slavery there were terrible instances of cruelty motivated by greed, pride, and worse. But is it really the case that religions do more harm than good? What about all the hospitals and schools and political movements driven by men and women of avowed religious commitment? How do we weigh all the indiscriminate acts of cruelty against all the apparently indiscriminate acts of kindness carried out in the name of religion? The nature of the comparison might be queried. It's a bit like asking if the world would be better without men. For the foreseeable future we are going to have both religions and men, so a better way forward is to ask how we can organize matters so that both contribute to human flourishing.

The Yale theologian Miroslav Volf has written how world religions, despite their malfunctions, remain one of our most potent sources of moral motivation and contain within them profoundly evocative accounts of human flourishing. He explains how

> globalization stands in need of the visions of flourishing that world religions offer... globalization and religions, as well as religions among themselves, need not clash violently but have internal resources to interact constructively and contribute to each other's betterment... I have attempted to identify the unity of meaning and pleasure as a wellspring of flourishing, a source of personal contentment, global solidarity, and common care for our planet that our globalized world needs and religions can foster... The idea can be expressed simply: the right kind of love for the right kind of God bathes our world in the light of transcendent glory and turns it into a theatre of joy.[21]

Professor Volf is intending to devote the bulk of his future work to fleshing out that claim and the vision of the good life which can be built on it. His analysis illustrates how religions can play a particular role in reconciliation, which is one of the three priorities of the present Archbishop of Canterbury, and was one of the great contributions of a former Archbishop of Cape Town.

On 27 April 1994 the first election in South Africa was held in which citizens of all racial groups (as they were then defined) could vote. Following the transition to majority rule, the issue had to be addressed of the widespread atrocities which had been committed under apartheid. How should that be undertaken? Should there be Nuremberg-style trials for gross violation of human rights? What about justice and reparations? Under the chairmanship of Desmond Tutu, Archbishop of Cape Town, the Truth and Reconciliation Commission was established (Figure 9.3). For applicants to be granted amnesty, four conditions had to be satisfied: the act for which amnesty was sought must have happened between 1960 and 1994; it must have been politically motivated; the applicant must make full disclosure of the facts; and the means must be proportional to the objective. This last meant in practice that the more serious the offence, the more public should be the hearing—this could have painful consequences for some whose families and friends thereby learned for the first time about what they had done.

Central to its work was Archbishop Tutu's Christian understanding of forgiveness and reconciliation.[22] He argued—successfully—that it should not be necessary for the perpetrator to express contrition or remorse in order to gain amnesty, but they had to accept responsibility. Victims of gross human rights violations gave statements about their experiences. The perpetrators had to give testimony and request amnesty from both civil and criminal prosecution. Amnesty was granted only to those who pleaded guilty; it was not given to those who claimed to be innocent. That was why it was not granted to the police officers who played a part in the death of Steve Biko, because they claimed that they were simply retaliating for his inexplicable conduct in attacking them. Amnesty was available only to those who accepted responsibility for their actions.

Archbishop Tutu concludes his account of the Truth and Reconciliation Commission with a profound statement of his motivating belief:

> God does have a sense of humour. Who in their right mind could ever
> have imagined South Africa to be an example of anything but awfulness;

Figure 9.3 The South African Truth and Reconciliation Commission was inspired by Christian ideals of forgiveness and reconciliation.

of how *not* to order a nation's race relations and its governance? We South Africans were the unlikeliest lot and that is precisely why God has chosen us...God wants us to succeed not for our glory and aggrandisement but for the sake of God's world. God wants to show that there is life after conflict and repression—that because of forgiveness there is a future.[23]

The Commission is generally held to have been successful.[24] Desmond Tutu's work was recognized by his being awarded both the Nobel Peace Prize and the Templeton Prize.

But is it true?

When John Betjeman asked (in Chapter 5), 'And is it true?', he was referring specifically to the Christian celebration of the birth of Jesus. Here, we broaden out the question (in so far as one can put it succinctly) to 'Are religions correct when they point to a God or something that lies beyond the natural?'

A wide range of arguments have been used in attempts to prove that God does or does not exist. (Spoiler alert—we are not about to resolve the question here.) Anselm's argument, which we introduced in Chapter 4, was not developed in the context of postmodern scepticism. Rather it was part of a movement in an era of widespread religious commitment to show that Christian belief could stand up to the intellectual tsunami of Greek philosophy as it was recovered through Arabic translations.

Enquiring about God's presence beyond the natural depends on what one means by God. In pantheism God and the Universe are one, so that in studying the Universe through the natural and social sciences one is studying God. A pantheist therefore rejects the notion that there is a God beyond the natural and considers it nonsense to ask if God is personal and whether one can have a relationship with God—one can have a relationship with God in the same sense as one can have a relationship with all of reality. Indeed, the very question 'Does God exist?' is questionable, because anyone asking that question is likely to have a concept of God that is less than God.[25] A professor of physics at Oxford expressed it rather challengingly:

> I would say that the question 'Does God exist?', to me, has already gone wrong, because it is like one of those paradoxical statements such as 'this statement is false'. If I hear the question, then I am ready to answer 'yes' in so far as the word 'God' really refers to the One who has that title, but the very fact that the questioner is asking alerts me to the fact that they are not truly referring to God. They are using the same word for some other notion, some sort of something which could possibly not exist. So the whatever-it-is that they are asking about clearly is not God. They are obviously in a muddle. How should I reply then? The best reply I can offer is something like, 'Let's talk further'. Or I might say something more provocative, such as: 'that's like asking, "Are perfect numbers composite?" ' (The fact that you need to ask shows that you don't know what perfect numbers are.) Or I might even say, 'no', because it sounds like what they mean by 'God' in their question will turn out to be all wrong.[26]

In mainstream religions, one of the oldest arguments about the existence of God is the cosmological argument. Essentially, this is a family of arguments all to do with first causes. The argument is an instance of natural theology in that it proceeds from observations of nature to conclusions about the existence of 'a first or sustaining cause, a necessary being, an unmoved mover, or a personal being (God) exists that caused and/or sustains the universe.'[27] The argument goes back at least to Plato and has historically been important in Judaism,[28] Christianity,[29] and Islam.[30]

A related argument seeks to establish from the existence of life on Earth whether there is a benevolent creator. We do not know how life began, but at some point we may find out. Suppose that it turns out to be exceedingly improbable that life began at all, perhaps with growing evidence of the absence of life anywhere else in the Universe. Aha, a theist says, that shows that God must have intervened to create life, and 'he' must be a good God to have done so. Ah no, replies an atheist, a good god would have created a universe in which life was certain to arise. Then let us suppose that life turns out to be highly probable, perhaps because evidence of life is found on many exoplanets. You see, cries the atheist, there is no need of god to create life. On the contrary, responds the theist, that shows the goodness of God in creating a universe so conducive to life. Such probabilistic gymnastics are frustratingly inconclusive.[31]

The cosmological argument may seem to be something of a period piece. We are not among those who see science as being in conflict with religion (quite the reverse) but few people nowadays seek to argue from features of the natural world to something (or someone) whom one maintains is outside of it. This is emotionally distinct from the experience of awe and wonder at the natural world, and intellectually distinct from someone who has a belief in God trying to understand something of God's nature from the features of the natural world. Michael has elsewhere argued how these features suggest—though one hesitates to ascribe attributes to God—that God is patient, encourages diversity, and takes risks.[32]

Related to the cosmological argument is the more familiar argument from design. The most famous exposition of the argument was given by Paley in 1802.[33] Paley imagined a person crossing a heath who chanced upon a watch. The complexity and evident purpose in the design of the watch, Paley argued, leads us to conclude that the watch had a maker. Paley asserted that what is true for a watch is true for as wonderful an organ as the human eye. Paley wasn't aware of it but his argument from design had already been refuted (in most people's eyes) by the philosopher David Hume some 23 years earlier.[34] Hume's refutation is complicated as he sees two versions of the argument from design, one to do with the regularity in the natural world, the other to do with adaptation. Essentially, Hume rejects analogies between the artificial world (Paley's watch) and the natural world (Paley's eye).

However, what did for Paley's version of the argument from design (as far as most people are concerned) wasn't the Humean critique but the Darwin–Wallace theory of natural selection. Darwin himself admitted that the notion that natural selection could account for the evolution of the human eye seemed 'absurd in the highest possible degree'[35] but then went on to argue that it could. Most people have come to accept Darwin's arguments but counter-arguments persist. The Anglican Bishop Hugh Montefiore cited a large number of biological observations which he felt could not satisfactorily be explained by natural selection. It must be said that to a scientist who is also a Christian, reading much of Montefiore's book is deeply embarrassing. To give one example:

> As for camouflage, this is not always easily explicable on neo-Darwinian premises. If polar bears are dominant in the Arctic, then there would seem to have been no need for them to evolve a white-coloured form of camouflage.[36]

As Richard Dawkins cruelly, but accurately, noted:

> This should be translated: I personally, off the top of my head sitting in my study, having never visited the Arctic, never having seen a polar bear in the wild, and having been educated in classical literature and theology, have not so far managed to think of a reason why polar bears might bene-fit from being white.[37]

Polar bears are probably white so as to avoid detection by their prey. They rely on surprise, frequently hunting seals that are resting on ice (Figure 9.4).

Figure 9.4 The adaptations shown by organisms appear to lend support to the argument from design for the existence of God but few scientists accept this argument.

A recent mathematical attempt to explore whether the existence of God can be investigated logically comes from the American political scientist Steven J. Brams. He had previously made a name for himself with a solution to the problem of sharing a cake in such a way that no one need feel envious of anyone else's share. For two people the solution has been known since ancient times; it can be found in the Hebrew bible following the arrival of Abraham and Lot in Canaan.[38] Brams and others extended this to more than two parties. Moving from cakes to the existence of God, Brams used game theory to explore how a superior being who possessed supernatural qualities of omniscience, omnipotence, immortality, and incomprehensibility would act differently from us, and whether these differences would be knowable.[39] It seems unlikely that many people have been persuaded by game theory to change their belief about the existence of God.

We could continue to look at the various arguments for God's existence and the ones against God's existence—such as the problem of evil. Few people seem to find such arguments convincing or, even if they find them intellectually convincing, find that such arguments cause them to change their mind—either to come to faith or to lose their faith. In part this may be because of too narrow a view of empirical evidence, forgetting 'that the task of history is a necessary, but normally absent, ingredient in "natural theology" '[40]. It may also be because, however much humans pride themselves on decision-making through rational processes, it is integral to how humans behave to make choices, especially political and religious allegiances, on other grounds than reason alone.[41]

Francis Spufford, whom we quoted in Chapter 3, dispenses with the apologetic tradition in which a defence is erected for religious faith. Rather, he simply maintains that religious faith for him is built up and sustained by emotions:

> And so the argument about whether the ideas are true or not, which is the argument people mostly expect to have about religion, is also secondary for me. No. I can't prove it. I don't know that any of it is true. I don't know if there's a God. (And neither do you, and neither does Professor Dawkins, and neither does anybody. It isn't the kind of thing you can know. It isn't a knowable item.) But then, like every human being, I am not in the habit of entertaining only the emotions I can prove. I'd be an unrecognisable oddity if I did.[42]

Spufford's argument has some similarities with an earlier one expressed by Alvin Plantinga with less passion but greater philosophical underpinning. In what is now known, thanks to Plantinga, as 'reformed epistemology', the central claim is made that religious belief can be rational without any appeal to evidence or even argument.[43] Plantinga therefore argues that people can know God as a 'basic belief'.[44] Plantinga is not claiming to prove the existence of God, let alone the veracity of the specifics of the Christian faith such as the doctrine of the Trinity. Rather, he claims that these beliefs are not irrational. The burden of proof then shifts to an opponent, who needs to demonstrate convincingly the falsity of such beliefs.[45] Plantinga goes on to argue that Christian beliefs are reliably formed in Christians through the guidance of the Holy Spirit and by the operation of a *sensus divinatus*—a sense of the divine.

We end this section by returning to our spoiler alert at its beginning: we cannot prove whether or not God exists and we rather doubt that such proofs are available in the sense that the term is used in mathematics or the sciences. In earlier times, the word 'proof' meant 'test'—as in a knight proving his valour: 'She dropped her glove, to prove his love, then looked at him and smiled.'[46] If anything or anyone is being tested by life in this world it is us rather than a divine being. In any event, religious faith has a subjective element that goes beyond intellectual proof. If you like, it is a choice, or a commitment. And it has huge consequences, not least for hope about what happens after the end of a flourishing life.

Life after death

What happens to us once we die? Some people think that near-death experiences provide strong (even conclusive) evidence for life after death, although we disagree. As we argued in Chapter 4, there may well be naturalistic explanations for these experiences. Some people think one can prove that there is no life after death; for example, the scientist Peter Atkins has argued that science disproves the possibility,[47] but this seems to us to be a category error. The question as to whether there is life after death or not simply lies outwith science.

Religions vary greatly in their views about life after death. Some religions—always remembering the great range of beliefs within religions—do not envisage any kind of existence after death, but most do. A common belief is in reincarnation—the belief that after we die, we start a new life in a different body. Reincarnation is found in Buddhism, Hinduism (itself an aggregate of different religions), Jainism, and Sikhism as well as in a number of ancient religions.

In Buddhism, reincarnation (also known as rebirth or metempsychosis) occurs as part of the cycle of death and rebirth. Rebirth is not necessarily as a human or even an animal—it can take place in the heavenly realm or one can be reborn as a demigod, a ghost, or a resident of hell. Which of these is one's fate is determined by *karma*—one's actions in one's most recent and one's previous lives. Ultimately, this endless cycle is meaningless, so one seeks for release and the attainment of *nirvana*, attained when one no longer manifests greed, hate, or ignorance.

In some understandings, nirvana is equated with *anatta* (non-self). The doctrine of anatta is a key difference between Buddhism and the Abrahamic faiths (Judaism, Christianity, and Islam). In Buddhism there is no self, and belief that there is self only causes suffering. Nevertheless, this belief in non-self is able to coexist with a belief in rebirth, though not rebirth of a soul that survives this life and is found again in the next.

In the mainstream understandings of the Abrahamic faiths, there is no reincarnation. Instead, as the author of Hebrews puts it 'humans have to die once, and after that comes judgment'.[48] At the Day of Judgement (also known as the Last Judgement), people are traditionally divided into those who go to heaven and those who go to hell—as depicted in many wonderful mediaeval paintings, such as Figure 9.5, that typically show the righteous entering heaven and the ungodly being carried off by devils to eternal damnation. Artistic depictions of hell (Bosch, *The Garden of Earthly Delights*; Bruegel, *Dulle Griet*; Jake and Dinos Chapman, *Fucking Hell*; Van Eyck, *The Last Judgment*) are generally more memorable than ones of heaven—rather as many actors prefer to play the villain than the saint.

This binary classification of all of humanity into 'eternally saved' and 'eternally lost' has caused theologians and others to feel uncomfortable. It is a problem that none of us seem good enough to go straight to heaven. A solution was to devise the notion of purgatory. In purgatory, we can be purified so that we are eventually capable of entering heaven. Purgatory then came to be seen as having a number of levels, the lowest being nearest to hell, through which one gradually ascended.

Whether or not one accepts the doctrine of purgatory—and many Christians, especially in the Protestant traditions, do not—there is a stark problem about hell. Why do some people go to hell and others

Figure 9.5 Stefan Lochner's *The Last Judgement* (*c*.1435).

don't? Traditionally, in Christianity, the answer has been that heaven is open to all who repent of their sins and confess that Jesus Christ is Lord but many do not take up this promise. For those who find this less than satisfactory the doctrine of universalism—that all end up in heaven—resurfaces repeatedly in Christian history.

A careful attempt to provide a theological answer to the question of what happens to us when we die was provided by John Hick. He attempted to provide a global theology of death by drawing on both Western and Eastern traditions without constructing a global religion. Hick's tentative hypothesis is that the state immediately after we die is 'subjective and dream-like'.[49] It can therefore be whatever the person who has died expects: 'the devoted Christian may find herself before the throne of final judgment, while the secularist might have a dream-like experience largely continuous with her earthly life'.[50] Hick sees life as a continuous process of soul-making and holds that most of us have not completed that process before we die. Our present life is therefore seen as 'the first of a series of limited phases of existence, each bounded by its own "death" '.[51] Hick espouses a version of universalism and ends his book by expressing the hope:

What Christians call the Mystical Body of Christ, with the life of God, and Hindus the universal Atman which we all are, and Mahayana Buddhists the self-transcending unity in the Dharma Body of the Buddha, consists of the wholeness of ultimately perfected humanity beyond the existence of separate egos.[52]

When we proposed this book to Oxford University Press, our editor asked us, as he asks every author, 'Who is this book for?' We gave the most accurate answer we could but there was one thing we could be certain of about all our readers which we understandably neglected to include in our proposal. All of them will die. That we all accept that, does not take away the pain of death and what all too often feels like the wrongness of death. There is often no good time to die, and no good way to die. There is what John Donne called the pride of death:

> Death, be not proud, though some have called thee
> Mighty and dreadful, for thou art not so;
> For those whom thou think'st thou dost overthrow
> Die not, poor Death, nor yet canst thou kill me.
> From rest and sleep, which but thy pictures be,
> Much pleasure; then from thee much more must flow,
> And soonest our best men with thee do go,
> Rest of their bones, and soul's delivery.
> Thou art slave to fate, chance, kings, and desperate men,
> And dost with poison, war, and sickness dwell,
> And poppy or charms can make us sleep as well
> And better than thy stroke; why swell'st thou then?
> One short sleep past, we wake eternally
> And death shall be no more; Death, thou shalt die.[53]

Despite the pride of death, the Christian hope is that death has lost its sting, as Handel so movingly set Paul's rhetorical question to music in *The Messiah*. Andrew's grandfather, in one of his hymns, described how 'by death destroying death, Christ opened wide life's gate.'[54]

It is hardly surprising that for those who believe in an eternal after-life, it makes a profound difference to their understanding of this life and what makes for human flourishing. As Paul put it: 'If it's only for this present life that we have put our hope in the Messiah, we are the most pitiable members of the human race.'[55] But if this life is indeed a short preparation for a world to come, then one might predict that people who believe in heaven are more positive about life than people who don't. Conversely, people who believe in hell might be more negative about life than people who don't. Belief in heaven does not necessarily equate with belief in hell. A recent study used this asymmetry to see whether people are more satisfied with life if they are more likely to believe in heaven than in hell.[56] The answer is that they are (Figure 9.6). Plenty of less quantitative thinking explains why the Christian hope of eternal life contributes to human flourishing. Once again, that does not prove that it is true, but it does suggest that it is worth asking, 'What are we waiting for? And what are we going to do about it in the meantime?'[57]

David Brooks has suggested that there are two sets of virtues, the résumé virtues and the eulogy virtues:

> The résumé virtues are the skills you bring to the marketplace. The eulogy virtues are the ones that are talked about at your funeral— whether you were kind, brave, honest or faithful. Were you capable of deep love?[58]

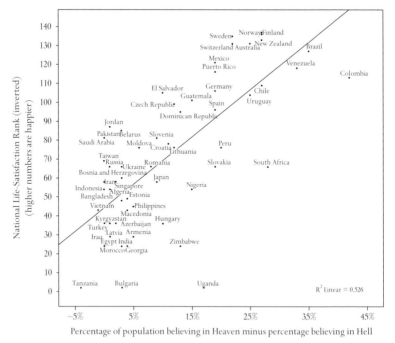

Figure 9.6 People who live in countries where the population is more likely to believe in heaven than in hell have greater life satisfaction.

There are many factors that can help us develop such eulogy virtues. One of these may be a belief that the person we are, what we do with our lives, and how we relate to others has consequences not just for this life, but for eternity.

Religiosity and human flourishing

It is not straightforward to measure either religiosity, which encompasses belief, practice, and experience, or human flourishing, which we have argued has material, relational, and transcendental dimensions. Even if we can, a correlation between the two does not necessarily imply causation. A positive correlation between quantitative measures of religiosity and human flourishing might be because being more religious tends to enhance flourishing, or the causal relationship could be the other way, or it might be that something else is causing both religiosity and flourishing.

A classic example of an indirect reason for a positive correlation was a spoof announcement in Australia that described a positive relationship between ice cream consumption in the coastal town of Port Bull over three years and the incidence of shark attacks.[59] The purported lead scientist, Dr John McQuaig, from the fictitious University of Wroxton, was reported as saying, 'This sheds intriguing light on the effects of diet on humans' appeal to carnivores, suggesting that saturated fat and dairy may be particularly attractive to sharks.' The report of the research findings went on to state:

> While not usually man-eaters, sharks require highly-caloric food to power their muscular bodies. As fat contains more calories per gram than protein or carbohydrates, it is believed that humans with higher levels of body-fat would be more attractive to sharks.
>
> 'Ice cream itself has a lot of fat,' says Dr. Anita Bath, another researcher from the University of Wroxton.

The point is that both ice cream consumption and shark attacks are more likely in the summer when the weather is hot and people buy ice creams and, quite independently, are presumably more likely to go swimming in shark-infested waters.

The problems with deducing causation from correlation are a major reason why many researchers use random controlled trials where individuals are randomly allocated to a treatment group and a control group. Any differences in outcomes can then be attributed to the treatment. This works quite well for some sorts of research, like trials of drugs—though even here patients sometimes correctly work out whether they are in the treatment or control group and then behave differently depending on which group they are in (e.g. seeking treatments from outside the trial if they conclude that they are in the control group). Somewhat remarkably, it is often the case that those who know that they are receiving a placebo experience a benefit over those who are not receiving any treatment at all (Figure 9.7). But random controlled trials don't really work when we are interested in the effects of religiosity on flourishing. A researcher can't really control variables like religious faith and religious experience, though they could, in principle, get people to agree to pray more or less or go to places of worship more or less.

Figure 9.7 Placebos or extra-strength placebos?

Allowing for all the caution with which quantitative findings on the relationship between religion and flourishing should be treated, what do the data tell us? The most robust studies tend to use data on religious participation, mainly because such data are more objective and likely therefore to be more reliable than data on one's religious experiences or the strength of one's faith (person X may modestly but sincerely state that their faith is weak whereas to most observers, in so far as anyone can ascertain the strength of someone else's faith, person X seems a deeply religious believer). Early studies sometimes suggested really large effects of attending worship on subsequent mortality rates— ignoring the fact that if you are very ill (and thereby close to death) you may not be able to attend worship!

There have now been several dozen reasonably robust studies looking at religiosity and mortality, encompassing over 100,000 participants. The studies control for variables like age, sex, ethnicity, geographic region, baseline health, and socioeconomic status.[60] The take-home message is that attending services is associated with a reduction in mortality rates (i.e. chances of dying over a fixed period of time) of about 20–35 per cent, with women typically showing a greater effect than men. These improvements are found in countries with very different rates of religious observance and across a range of religions: Denmark, Finland, and the USA (Christian), Israel (Jewish), and Taiwan (Taoist and Buddhist).

What about religion and mental health? For instance, is religion associated with psychological delusions? Sigmund Freud thought so:

> A special importance attaches to the case in which [the] attempt to procure a certainty of happiness and a protection against suffering through a delusional remoulding of reality is made by a considerable number of people in common. The religions of mankind must be classed among the mass-delusions of this kind.[61]

Freud's views on religion changed over the course of his writing,[62] but not before he had convinced many followers that religion is generally associated with poor mental health. Quantitative studies in recent decades indicate that the reverse is typically the case. A recent US study of 66,000 female registered nurses and 43,000 male health care professionals found that attendance at religious services at least once per week was associated with a 68 per cent lower likelihood among women of death over the next 16 years from drugs, alcohol, or suicide (a huge difference), and a 33 per cent lower likelihood among men, compared with those who said they never attend religious services.[63] More generally, attending religious services is associated with lower levels of depression; high-quality studies indicate that the association may be causal.[64]

However, there is some evidence that the effect of attendance at religious services is less pronounced or even counterproductive in countries which restrict freedoms or in countries in which religious participation is less common. One study suggested that students in schools where their own religious affiliation was in the minority were more likely to attempt suicide or self-harm.[65] This raises the possibility that at least some of the generally positive effects of religion on health may be to do with social conformity.

In a highly cited article titled 'The Religion Paradox: If Religion Makes People Happy, Why Are So Many Dropping Out?', Diener and colleagues asked why people are rapidly leaving organized religion in economically developed nations where religious freedom is high.[66] After controlling for circumstances in both the United States and world samples, they found that religiosity is associated with slightly higher subjective measures of well-being across Buddhism, Christianity, Hinduism, and Islam. These associations were mediated by social support, feeling respected, and purpose or meaning in life. In countries where life is tougher (e.g. widespread hunger and low life expectancy), people are more likely to be highly religious. However, in countries where life is easier, people are less likely to be highly religious, *and* religious and nonreligious individuals report similar levels of subjective well-being.

In a subsequent article, Myers and Diener reported a further religious engagement paradox. If you compare more or less religious countries, or different American states, then religious engagement correlates *negatively* with measures of well-being. But if you compare individuals within a given country or state, especially where religion is more prevalent, then religious engagement correlates *positively* with measures of well-being. What is going on? Countries with high levels of religious engagement are often poorer, and the effect of religious engagement may be swamped by factors that detract from well-being, such as food shortage, poor health-care, and low standards of education. Controlling for these factors eliminates the negative correlation between religion and well-being.[67]

Religiosity is not invariably associated with good mental health. There are many examples of people who are what most of us would think of as excessively religious who do not enjoy good mental health. Nevertheless, the overall association between religion and health seems to be a positive one. Of course, human flourishing is about more than health or longevity. The story for the association between religion and such self-reported measures of human flourishing as life satisfaction or subjective well-being is much the same as for the association between religion and health. In high-quality studies which fully address the methodological issues which we raised above, human flourishing and religion go together.[68]

Hope springs eternal

Each Christmas Eve a Festival of Nine Lessons is broadcast live from the chapel of King's College Cambridge. The service was introduced in 1918, immediately following the end of World War 1, by Eric Milner-White, who subsequently worked with Andrew's grandfather.[69] The following year the order of service was revised to begin with the carol 'Once in Royal David's City'. In 1928 the BBC broadcast the service live. Andrew's father sang as a chorister in that broadcast, and in subsequent broadcasts, and later as an undergraduate choral scholar. The first verse of the opening carol is sung as an unaccompanied treble solo, which for many who hear it marks the start of Christmas. None of the choristers knows which of them will be chosen to sing the solo; only at the last minute does the Director of Music single out one of them and give him the opening note on a tuning fork. Andrew's father was once asked on a BBC interview whether he remembered being nervous. He replied that they were not so much thinking of the large audience, but rather about the music and the words:

> And our eyes at last shall see Him,
> Through His own redeeming love;
> For that child so dear and gentle,
> Is our Lord in heaven above,
> And He leads His children on,
> To the place where He is gone.

How well does this capture the contribution of religious hope to human flourishing? This sense of now we are here, and then we shall be there? If time's arrow points us towards the next life, then that should provide an ethical incentive for how we live now by giving an eternal dimension to David Brooks' eulogy virtues. It provides a reason not to lay up treasure for ourselves on earth, where it can be eroded by moths and rust and financial crashes, but to lay it up in heaven where it will not be susceptible to the vicissitudes of fortune. In the hall of the Worshipful Company of Clothworkers, a mediaeval guild established to maintain the technical and professional standards of what was then a major national industry, there is a stained-glass window which includes a moth as the patron insect of the trade, because it helps to promote obsolescence and thereby maintain sales.

Like any good incentive, the hope of heaven can be abused. 'Pie in the sky when you die', as the rhyming cliché expresses it, can be an excuse for social inaction now. It can lead to the caricature of the saint who is so heavenly minded that they are no earthly use. At its worst it can be used to justify oppression and injustice.[70] The wonderful gospel songs, with their hope of the sweet chariot coming for to carry me home and my home over the deep river Jordan, which Andrew remembers his father singing with conviction, were born of terrible suffering and deprivation. But *abusus non tollit usum* (the misuse of something does not eliminate the possibility of its correct use), and at its best the hope of heaven can inspire the highest ethical standards.

However, that is not the whole story. It is not even the best part of the story. The direction of travel in the opening carol of the Festival of Nine Lessons and Carols, like the carol with the aspiration to 'fit us for heaven, to live with thee there', is from earth to heaven, from this life to the next. But in one of the best-known prayers of all time, one which is said every Sunday in churches all around the world, the direction of travel is the other way. Your kingdom come follows a desire that what is already the case in heaven should become the case on earth.[71] Everything that is hoped for beyond this material world should begin to become true in this world.[72] Those asking for that have their part to play in bringing it about. The prayer is named after the man who said that he came so that everyone could have life full to overflowing.[73] He even gave knowing God as his own definition of what he meant by human flourishing.[74]

Human flourishing this side of the grave is to be carried out in the everyday world of our existence, with its material, relational, and transcendent aspects. The principal intellectual resource which we have for engaging with the material world is science. Advances in science have a habit of leading to technologies which at their best provide instruments for promoting human flourishing.[75] But it can seem, as an Oxford philosopher recently expressed it, that 'There is a long-distance race on between humanity's technological capability, which is like a stallion galloping across the fields, and humanity's wisdom, which is more like a foal on unsteady legs.'[76] We turn next in the application of our foundations of human flourishing to case studies in emerging technologies.

Notes

1 Woodhead, L. and Catto. R. (Eds) (2012) *Religion and Change in Modern Britain*, London: Routledge.

2 Brierley, P. (Ed.) (2006) *UK Christian Handbook: Religious Trends, 6, 2006/7, Pulling out of the Nosedive*, London: Christian Research.

3 Faith Survey (2020) Christianity in the UK: Measuring the Christian population in the UK. Available at https://faithsurvey.co.uk/uk-christianity.html.

4 Ali, S. et al. (2015) *British Muslims in Numbers*, London: The Muslim Council of Britain. Available at http://www.mcb.org.uk/wp-content/uploads/2015/02/MCBCensusReport_2015.pdf.

5 Office for National Statistics (2018) Muslim population in the UK. Available at https://www.ons.gov.uk/aboutus/transparencyandgovernance/freedomofinformationfoi/muslimpopulationintheuk/.

6 James, W. (1902) *The Varieties of Religious Experience: A Study in Human Nature, Being the Gifford Lectures on Natural Religion Delivered at Edinburgh in 1901–1902*, New York: Longmans, Green.

7 Pearce, L. D., Hayward, G. M. and Pearlman, J. A. (2017) Measuring five dimensions of religiosity across adolescence, *Review of Religious Research* **59**(3), 367–93.

8 Bonde, J.-P. (2009) *The Lisbon Treaty: The Readable Version*, third edn, Foundation for EU Democracy. Available at http://en.euabc.com/upload/books/lisbon-treaty-3edition.pdf.

9 Siedentop, L. (2014) *Inventing the Individual*, London: Allen Lane.

10 Cupitt, D. (1984) *The Sea of Faith*, London: BBC Books.

11 Don Cupitt official website, About non-realism. Available at https://www.doncupitt.com/non-realism.

12 Cox, H. (1965/2013) *The Secular City*, Princeton, NJ: Princeton University Press.

13 Taylor, C. (2007) *A Secular Age*, Cambridge, MA: The Belknap Press of Harvard University Press, p. 3.

14 Dawkins, R. (2006) *The God Delusion*, London: Bantam Press.

15 Hitchens, C. (2007) *God Is Not Great: The Case against Religion*, London: Atlantic Books.

16 Berger, P. (1999) *The Desecularization of the World: Resurgent Religion and World Politics*, Grand Rapids, MI: Eerdmens.

17 Tearfund (2020) Many Brits look to faith during lockdown, 3 May. Available at https://www.tearfund.org/en/media/press_releases/many_brits_look_to_faith_during_lockdown/.

18 Hitchens (2007) *God Is Not Great*, p. 18.

19 Jones, J. W. (2008) *Blood That Cries out from the Earth: The Psychology of Religious Terrorism*, Oxford: Oxford University Press.

20 Neumann, P. R. (2010) *Prisons and Terrorism Radicalisation and De-radicalisation in 15 Countries*, London: International Centre for the Study of Radicalisation and Political Violence. Available at https://www.clingendael.org/sites/default/files/pdfs/Prisons-and-terrorism-15-countries.pdf.

21 Volf, M. (2015) *Flourishing: Why We Need Religion in a Globalized World*, New Haven, CT: Yale University Press.

22 Shore, M. (2008) Christianity and justice in the South African Truth and Reconciliation Commission: A case study in religious conflict resolution, *Political Theology* **9**(2), 161–78.

23 Tutu, D. (1999) *No Future without Forgiveness*, Ryder: Random House.

24 Fullard, M. and Rousseau, N. (2009) Truth-telling, identities and power in South Africa and Guatemala. Research brief. International Center for Transitional Justice, Research Unit. Available at https://www.ictj.org/sites/default/files/ICTJ-Identities-TruthCommissions-ResearchBrief-2009-English.pdf.

25 Briggs, A., Halvorson, H. and Steane, A. (2018) *It Keeps Me Seeking: The Invitation from Science, Philosophy, and Religion*, Oxford: Oxford University Press, p. 7.

26 Ibid., pp. 17–18.

27 Reichenbach, B. (2017) Cosmological argument, *The Stanford Encyclopedia of Philosophy*, 11 October. Available at https://plato.stanford.edu/archives/fall2019/entries/cosmological-argument/.

28 Wolfson, H. A. (1924) Notes on proofs of the existence of God in Jewish philosophy, *Hebrew Union College Annual* **1**, 575–96.

29 Aquinas (1265–74) *Summa Theologiae*.

30 Craig, W. L. (1979) *The Kalām Cosmological Argument*, New York: Barnes & Noble.

31 Briggs, Halvorson, and Steane (2018) *It Keeps Me Seeking*, Chapter 11.

32 Reiss, M. J. (1993) The argument from design, *Modern Churchman* **34**, 105–10.

33 Paley, W. (1802) *Natural Theology, or Evidences of the Existence and Attributes of the Deity collected from the Appearances of Nature*, London: J. Vincent.

34 Hume, D. (1779) *Dialogues Concerning Natural Religion*, London. The dialogues were only published posthumously and initially identified neither the author nor the publisher.

35 Darwin, C. (1859) *On the Origin of Species by Means of Natural Selection, or the Preservation of Favoured Races in the Struggle for Life*, London: John Murray, p. 143

36 Montefiore, H. (1985) *The Probability of God*, London: SCM, p. 82.

37 Dawkins, R. (1988) *The Blind Watchmaker*, Harmondsworth, UK: Penguin, p. 38.

38 Genesis 13: 8–12.

39 Brams, S. J. (1983) *Superior Beings: If They Exist, How Would We Know?* New York: Springer-Verlag.

40 Wright, N. T. (2019) *History and Eschatology: Jesus and the Promise of Natural Theology*, London: SPCK, p. 74.

41 Haidt, J. (2012) *The Righteous Mind: Why Good People are Divided by Politics and Religion*, London: Allen Lane.

42 Spufford, F. (2012) *Unapologetic*, London: Faber and Faber, p. 21.

43 Bolos, A. and Scott, K. (2020) Reformed epistemology, Internet Encyclopedia of Philosophy. Available at https://www.iep.utm.edu/ref-epis/.

44 Plantinga, A. (2000) *Warranted Christian Belief*, New York: Oxford University Press.

45 For an example of someone who is not persuaded by Plantinga's arguments, see Koons, J. R. (2011) Plantinga on properly basic belief in God: Lessons from the epistemology of perception, *The Philosophical Quarterly* **61**(245), 839–50.

46 Hunt, L. (1784–1859) The glove and the lions.

47 Atkins, P. (2011) *On Being: A Scientist's Exploration of the Great Questions of Existence*, Oxford: Oxford University Press.

48 Hebrews 9: 27.

49 Hick, J. (1976/1985) *Death and Eternal Life*, Houndmills, UK: Macmillan, p. 416.

50 Cramer, D. C. (2020) John Hick (1922–2012), Internet Encyclopedia of Philosophy. Available at https://www.iep.utm.edu/hick/#SH3c.

51 Hick (1976/1985) *Death and Eternal Life*, p. 408.

52 Ibid., p. 464.

53 Donne, J. (1571–1631) Death, be not proud (Holy Sonnet 10). Available at https://poets.org/poem/death-be-not-proud-holy-sonnet-10.

54 Briggs, G. W. (1875–1959) Now is eternal life. Available at http://brooke.churchnet.co/res/Covid-19/Words_for_Now_is_eternal_life.pdf.

55 1 Corinthians 15: 19.

56 Shariff, A. F. and Aknin, L. B. (2014) The emotional toll of hell: Cross-national and experimental evidence for the negative well-being effects of hell beliefs, *PLoS ONE* **9**(1): e85251.

57 Wright, N. T. (2007) *Surprised by Hope*, London: SPCK, p. xi.

58 Brooks. D. (2015) The moral bucket list, *The New York Times*, 11 April. Available at https://www.nytimes.com/2015/04/12/opinion/sunday/david-brooks-the-moral-bucket-list.html?_r=0; Brooks. D. (2016) *The Road to Character*, London: Penguin.

59 Arvik (2011) Ice cream consumption linked to shark attacks, Intergalactic Writers Inc. Available at https://intergalacticwritersinc.wordpress.com/2011/03/28/ice-cream-consumption-linked-to-shark-attacks/.

60 VanderWeele, T. J. (2017) Religion and health: A synthesis, in J. R. Peteet and M. J. Balboni (Eds), *Spirituality and Religion within the Culture of Medicine: From Evidence to Practice*, New York: Oxford University Press, pp. 357–401.

61 Freud, S. (1930/1962) *Civilization and its Discontents*, trans. James Strachey, New York: W. W. Norton, p. 28.

62 Thornton, S. (2020) Sigmund Freud: Religion, Internet Encyclopedia of Philosophy. Available at https://www.iep.utm.edu/freud-r/.

63 Chen, Y., Koh, H. K., Kawachi, I., Botticelli, M. and VanderWeele, T. J. (2020) Religious service attendance and deaths related to drugs, alcohol, and suicide among US health care professionals, *JAMA Psychiatry* 77(7), 737–44.

64 VanderWeele (2017) Religion and health: a synthesis.

65 Ibid.

66 Diener, E. D., Tay, L. and Myers, D. G. (2011) The religion paradox: If religion makes people happy, why are so many dropping out?, *Journal of Personality and Social Psychology* **101**(6), 1278–90.

67 Myers, D. G. and Diener, E. (2018) The scientific pursuit of happiness, *Perspectives on Psychological Science* **13**, 218–25.

68 VanderWeele (2017) Religion and health: a synthesis.

69 Milner-White, E. and Briggs, G. W. (1966) *Daily Prayer*, Harmondsworth, UK: Penguin.

70 Lindsay, B. (2019) *We Need To Talk About Race: Understanding the Black Experience in White Majority Churches*, London: SPCK, pp. 38–41.

71 Matthew 6: 9–13.

72 Wright (2007) *Surprised by Hope*.

73 John 10: 10.

74 John 17: 3.

75 Broers, A. (2005) *The Triumph of Technology: The BBC Reith Lectures 2005*, Cambridge, UK: Cambridge University Press.

76 Future of Humanity Institute (2018) £13.3m boost for Future of Humanity Institute. Available at https://www.fhi.ox.ac.uk/grant-announcement/#:~:text=Foundingper cent20Directorper cent20ofper cent20theper cent20FHI, appleper cent20specificallyper cent20forper cent20theper cent20foal.per centE2per cent80 per cent9D.

10

Human Flourishing in an Age of Technology

The word 'technology' comes from two Greek words: τέχνη, which roughly means 'craft', and λογία, which can be translated as 'treatise' or 'theory'. There is a virtuous circle between science and technology. Just as Anselm in the eleventh century described theology as *fides quaerens intellectum*—faith seeking understanding—so branches of science such as materials often seem to be *praxis quaerens intellectum*—practice seeking understanding.[1] The technology of steam power in the nineteenth century led to what is likely to remain one of the most robust branches of physics ever, thermodynamics. Sir Arthur Eddington expressed it memorably in a Gifford lecture:

> The law that entropy always increases holds, I think, the supreme position among the laws of Nature. If someone points out to you that your pet theory of the universe is in disagreement with Maxwell's equations—then so much the worse for Maxwell's equations. If it is found to be contradicted by observation—well, these experimentalists do bungle things sometimes. But if your theory is found to be against the Second Law of Thermodynamics I can give you no hope; there is nothing for it but to collapse in deepest humiliation.[2]

Technology in turn makes possible scientific discoveries. J. J. Thompson was able at the end of the nineteenth century in Cambridge to discover the electron because he had a high-quality vacuum. Over a hundred years later scientists at CERN in Geneva were enabled to discover the Higgs boson because of the extraordinarily sophisticated technology of the Linear Hadron Collider. In Andrew's laboratory at Oxford, industrially available dilution refrigerators make it possible routinely to perform experiments a fiftieth of a degree above absolute zero, which is colder than anywhere we know of outside a human laboratory. In turn, advances in science make possible improvements and innovations in technology, and so the virtuous circle goes round.

Figure 10.1   Neolithic stone and flint tools from Essex, England, c.2700–1800 BCE. Stone tools survive better than other technologies so our knowledge of early technology is dominated by them.

Most technologies do not preserve well in the fossil record, so our understanding of early technology is dominated by stone tools (Figure 10.1). The earliest of these survive from about 3.3 million years ago[3]—roughly a million years before the genus *Homo*, to which we belong, came into being. By subsequent standards, these tools look crude; they probably helped the individuals who used them to open nuts with thick shells. Gradually, finer tools appear in the fossil record. A great range of techniques were used in making them. For example, microliths dating from up to 35,000 years ago are pieces chipped off from a larger flint and then reworked. They are used as spear points or arrowheads, with up to 18 microliths in a spear or harpoon but only one or two for an arrow. Other techniques were used for making axes, chisels, and tools for cutting or grinding food and for polishing.

The gunflint industry, in which a spark from flint is used to fire a gun, survived into the middle of the twentieth century. Knives made from obsidian (a type of volcanic glass) are still used for certain types of surgery, as they are sharper than steel and cheaper than diamond knives. The fossil record for fire is not as clear as for stone tools. Our ancestors may have been using fire 1.7 million years ago; they certainly were 200,000 years ago.[4] Agriculture is only about 11,500 years old.[5] The wheel is about 5,500 years old.[6] Periods of human history are often denoted by the materials used: stone age; bronze age; iron age. The nineteenth century has been described as the steel age, and the twentieth as the silicon age.[7] The twenty-first century may mark a departure if it becomes dubbed the information age.

For many years, tools were thought to be one of the defining characteristics of humans. Jane Goodall's work with the chimpanzees of Gombe in Tanzania changed that.[8] In October 1960 she watched the male chimpanzee whom she called David Greybeard bend a twig, strip its leaves from it and use it to fish for termites in a termite mound. Nowadays, we know that quite a number of non-human animals, not all of them mammals, regularly use tools.

In an experiment reminiscent of Aesop's fable about a crow dropping stones into a jug to raise the water level so that it could drink, Caledonian crows were given a series of experiments designed to elucidate how they would solve problems involving fluid displacement.[9] The crows worked out that they should drop stones into a water-filled tube rather than a sand-filled tube, that floating objects were useless but sinking objects raised the water level, and that it was quicker to drop objects into a tube with a higher water level. There were some things that the crows failed to work out, such as that the water rises faster in narrow tubes than in wide tubes, and how to raise the water level in a tube that was too narrow to drop stones into but which had a hidden connection to a wide tube. The performance of the crows in the tasks in which they succeeded is comparable to children aged 5–7 years. The hidden connection problem is generally solved only by children older than that. Later experiments in the same laboratory used an arrangement where crows learned to obtain a reward by dropping heavy, but not light, objects into a food dispenser. They were then allowed to observe heavy and light objects hung on strings in a breeze created by an electric fan. Without any opportunity to handle the objects prior to the experiment, in 73 per cent of the trials the birds went for the heavy object, suggesting that they already knew about the behaviour of light and heavy objects in a wind.[10]

It thus seems that some animals not only use tools, which if you like you can call technology, but they also exhibit causal understanding, which if you like you can call science. It is not surprising that, like almost every other human metabolic, physiological, and neurological capacity, these capabilities are found in incipient form in animals. Nevertheless, humans use technology in a way that is not just quantitatively but qualitatively different from other species. This seems to be inseparable from the ability of humans to cooperate, over space and time, on projects with a scale and complexity that vastly exceed what any other species can do. Neil Armstrong could not have got to the Moon and back by himself. Technology and cooperation go together, and both of them are integral to human flourishing.

New technologies can be greeted with overenthusiasm or categorical rejection, sometimes with nerds and entrepreneurs cheering them on and cultural conservatives and ethicists crying 'Whoa!' Georgius Agricola was a sixteenth-century German mineralogist and metallurgist. His work was entirely based on his own first-hand study, and remained a definitive reference for over two centuries. He wrote of riches in general, and of metals in particular: 'For good men employ them for good, and to them they are useful. The wicked use them badly, and to them they are harmful.'[11]

John Browne was for twelve years CEO of one of Europe's largest oil companies. He quoted Agricola, and went on to summarize how every technology will generate its own set of consequences, intentional and unintentional, constructive and destructive:

> The same engineering that produces drones that deliver medicines to remote and disease-stricken communities also produces drones that are used in assassinations. Genetic engineering will cure some diseases, but it could also produce new pathogens. Opioid painkillers can relieve suffering, but can also cause addiction. Open communication and connectivity have allowed us to be expansive in our access to data and the use of it, but have also permanently weakened our ability to keep things private. Since the discovery and large-scale manufacture of penicillin, antibiotics have saved billions of lives, but their indiscriminate use has led to the appearance of drug-resistant bacteria, which if not eliminated, will cause great suffering. Hydrocarbon fuels have been the foundation of the great advances since the eighteenth century but using them produces greenhouse gases, which are dangerously altering the Earth's climate.[12]

The way that people choose to use a technology is what determines its impact on human flourishing. A general pattern, epitomized by the knitting frames and power looms that the Luddites feared would steal their livelihoods in the early nineteenth century, is for a wave of initial fear to be followed by a subsequent wave of new employment. This can run alongside what has become known as the Gartner Hype Cycle, which describes how technologies can progress through five stages of an innovation trigger, a peak of inflated expectations, a trough of disillusionment, a slope of enlightenment, leading to a plateau of productivity.[13] At its best, the plateau of productivity creates new opportunities for flourishing.

It is good that concerns are raised early in the development of a new technology, precisely because they serve as a counterbalance to uncritical endorsement (Gartner's peak of inflated expectations). Nevertheless, there are rarely conclusive reasons for categorically rejecting a new technology. Categorical rejection can have a number of roots—it may be that the new technology is thought to be 'unnatural' or that it cannot be proved to be safe. In one sense any technological development is unnatural by virtue of being new; equally, something that is new cannot be guaranteed to be safe precisely because it hasn't yet been around long enough to have its longer-term consequences evaluated. Michael used to sit on the UK Government's Advisory Committee on Novel Foods and Processes, the primary role of which was to advise ministers on matters of food safety. A running joke among members of the committee was that we would never have allowed potatoes to be introduced to the UK; our standards are much higher than when potatoes were first brought back from South America in the sixteenth century. In case you don't know, sprouting potatoes contain raised levels of the glycoalkaloids solanine and chaconine, both of which are natural toxins. The highest concentrations are found in the leaves, flowers, 'eyes', green skin, and sprouts. Ingesting even quite small amounts results in gastrointestinal symptoms such as vomiting, abdominal pain, and diarrhoea, though fatalities are infrequent. Similarly, despite rising safety standards, there are over a million road traffic deaths annually. If cars were invented today, would we permit them?

New technologies automate tasks, but they need not necessarily eliminate jobs. How many jobs in the twenty-first century would not exist if it were not for twentieth century developments in technology, from aviation to pharmaceuticals and computing? And how differently would we have had to cope with the coronavirus pandemic without digital networking technologies? It may, of course, be that the fear of losing one's job goes deeper, because of the associated loss of the dignity of a job which uses one's skill and competence.

A theory of what counts as a good life for us must nowadays include an explicit conception of how to live well with technologies: 'the invention of the bow and arrow afforded us the possibility of killing an animal from a safe distance—*or* doing the same to a human rival, a new affordance that changed the social and moral landscape.'[14] The philosopher Shannon Vallor, Past President of the Society for Philosophy and Technology, applies virtue ethics to technology, drawing on Aristotle, Confucian ethics (originating with Kongzi), and Buddhism. We can become more virtuous by engaging our will to avoid doing what we judge is not right and by getting into the habit of doing what we judge is right. To give a non-technological example, if I know from bitter experience that I have a tendency to drive home even if I have drunk more alcohol than I should, a virtuous course of action would be to get into the habit either of not drinking when out for an evening (when I will be driving myself home) or of relying on some other way of getting home. Virtue ethics provides a way of choosing what is right through the cultivation of our moral traits and capacities.

Issues of human flourishing need to be addressed early in the development of a new technology, just as safety should be considered early in the design of any new engineering project. This requires long-term thinking, supremely in educating children for the world of work in which they will make their way as adults.[15] As part of the Programme for the International Assessment of Adult Competencies (PIAAC), an OECD survey of literacy, numeracy and problem-solving skills was scathing in its findings. Back in 2017 only 13 per cent of workers used such skills on a daily basis with a proficiency demonstrably higher than computers. The report reckoned that even the best existing educational systems were unable to provide literacy, numeracy, and problem-solving skills needed to enable the majority of workers to compete with machines then, let alone the machines that were to come:

> There are no examples of education systems that prepare the vast majority of adults to perform better in the three PIAAC skills areas than the level that computers are close to reproducing. Although some education systems do better than others those differences are not large enough to help most of the population overtake computers with respect to PIAAC skills.[16]

In 2019 Andrew convened a dinner in the Palace of Westminster at which Andreas Schleicher, OECD Director of Education and Skills, spoke eloquently about these issues. He is well aware of the challenges, and is giving a global lead in addressing them. To what extent can we foresee the ways in which machine learning will impact on human flourishing?

AI and machine learning

In 1955 a young mathematician at Dartmouth College, USA, by the name of John McCarthy introduced the term artificial intelligence in a grant proposal to fund a two-month research project. He made no suggestion that machines might be able to think or understand, simply that they might be able to *simulate* aspects of intelligence:

> We propose that a 2 month, 10 man study of artificial intelligence be carried out during the summer of 1956 at Dartmouth College in Hanover, New Hampshire. The study is to proceed on the basis of the conjecture that every aspect of learning or any other feature of intelligence can in principle be so precisely described that a machine can be made to simulate it. An attempt will be made to find how to make machines use language, form abstractions and concepts, solve kinds of problems now reserved for humans, and improve themselves.[17]

It remains an open question whether machines ever will be aware that they have learnt something in the way that we are. Five years earlier the pioneer computer scientist Alan Turing had devised a test which came to be known after him.[18] He decided to replace the question 'Can machines think?' with a test in which the experimenter would ask questions via a keyboard, and from the printed answers seek to discern whether they came from a machine or a human. So far, no machine has fully passed the Turing test, although it has been suggested that the evasiveness of some politicians in answering questions might cause them to fail it.[19]

Thirty years later the philosopher John Searle produced his 'Chinese room' thought experiment to refute the possibility of what he called 'strong AI' which holds that 'the appropriately programmed computer really is a mind, in the sense that computers given the right programs can be literally said to understand and have other cognitive states'.[20] Searle imagines that AI has got to the point where a computer can be asked questions in Chinese and then, by virtue of its software program, produce answers that are as good as those that a human would produce. In other words, it passes the Turing test.

Searle then asks: does this show that the machine *understands* Chinese (strong AI) or merely *simulates* the ability to understand Chinese (weak AI)? Searle concludes that the thought experiment produces no evidence for strong AI for the following reason: imagine that Searle (who neither reads nor writes Chinese) is in a room and is given questions in Chinese along with a complete set of instructions (analogous to a computer program) as to how to answer in Chinese. Searle too would pass the Turing test but without understanding anything of the inputs or outputs.

A distinction is sometimes made between broad artificial general intelligence, AGI, and narrow artificial intelligence, AI. Humans, and to a remarkable extent some other animals, including many birds, can apply their intelligence to an extraordinary range of problems. AGI is a vision, some would say a dream or a nightmare, of machines that will be able to do that. If AGI could exceed human intelligence in most or all respects, then machines, or a machine, could take over the world. This is the stuff of both fiction and serious philosophical and technical speculation. It has been variously described as *Superintelligence,*[21] *The Technological Singularity,*[22] and *Life 3.0.*[23] It has been the subject of movies such as *Her* and *Ex Machina* and novels such as *Machines Like Me.*[24] The basic idea is the same in all of these: machine intelligence develops to the point where the machine is able to reset its own goals, and thereby pursue actions which may be to the detriment of its human creators. It is too early to say whether such AGI will ever happen, but we can be confident that it will not happen soon.

The fastest developing kind of AI is machine learning. Conventionally, computers were programmed to *do* a task, such as keeping the record of your bank account. Now, computers can be programmed to *learn to do* a task, such as verifying that a face in front of the camera is the same as the one in a passport. This is not AGI—the passport-checking machine cannot learn to boil an egg or tie its shoelaces—and in any case there is no widely agreed definition even of AI.[25] But machines are learning to outperform humans in a growing range of tasks that depend on intelligence of one kind or another in areas of life as diverse as manufacturing,[26] transport,[27] laboratory research,[28] and journalism. In 2020, around 27 journalists were sacked after Microsoft decided to replace them with artificial intelligence software.[29]

Andrew and his colleagues use machine learning in his laboratory in Oxford. They undertake research on solid state devices for quantum technologies such as quantum computing. They cool a semiconductor device with tiny regions created by nanoscale lithography to a temperature of 20 mK, and put one electron into each region, which may be only 1/1000 the diameter of a hair on your head. They then have to tune up the very delicate quantum states. Even for an experienced researcher this can take several hours. At the time of writing the 'machine' has learned how to tune their quantum devices in 12 minutes.

The students in his laboratory are now very reluctant to tune devices by hand. It is as if all your life you have been washing your shirts in the bathtub with a bar of soap. It may be tedious, but it is the only way to get your shirts clean, and you do it as cheerfully as you can . . . until one day you acquire a washing machine, so that all you have to do is put in the shirts and some detergent, shut the door, and press the switch. You come back two hours later, and your shirts are clean. You never want to go back to washing them in the bathtub with a bar of soap. And no one wants to go back to doing experiments without the machine. In Andrew's laboratory the machine decides what the next measurement will be.

Human intelligence enables us and our technologies (including software) to build on a cumulative history of human thought and action.[30] This applies as much to machine learning as to other human activities and technologies.[31] In 2016 AlphaGo beat eighteen-time world Go champion Lee Sedol by four game to one. Behind AlphaGo were all the coders and Go experts who worked to produce the software. Their combined intellectual effort was sufficiently significant to produce a paper in the prestigious journal *Nature*.[32] AlphaGo could not have come about without the remarkable capacity for human cooperation on a vast scale, enabled through the invention of long-lasting records such as writing and the capacity to accumulate both knowledge and ways of thinking and pass them on through education.

AlphaGo was awarded an honorary 9 dan—the highest grandmaster rank—by South Korea's Go Association. The citation on the certificate said it was given in recognition of AlphaGo's 'sincere efforts' to master Go's Taoist foundations and reach a level 'close to the territory of divinity'.[33] AlphaGo determined its next move based on the position on the board and deep learning previously acquired through extensive training, using both actual games from the whole history of Go and competitive games played by a copy of itself with a little randomness built in to avoid unfruitful repetition. AphaGo's successors no longer learn from human games—they simply play against themselves and rapidly improve through reinforcement learning.[34] They are sufficiently versatile to learn to play chess and other games as well.[35] The same team that created AlphaGo is now working on AlphaFold, to solve a grand challenge in biology that has been around for 50 years. AlphaFold promises to be able to calculate the shape and dimensions of each of the tens of thousands of proteins that make up the human body, thereby elucidating the role that healthy proteins play and accelerating the development of drugs that can attack proteins associated with diseases.[36]

Extraordinary advances are being made in the use of machine learning for health care. Machines can diagnose conditions such as breast cancer[37] and diabetic retinopathy[38] at least as accurately as experienced doctors. Researchers at Google Health and collaborators trained a deep-learning system for identifying breast cancer using screening mammograms from large UK and US data sets.[39] The machine outperformed every radiologist in the study. The website of the Royal College of Radiologists states, 'Clinical radiology is one of the most exciting and rapidly advancing specialties within medicine.'[40] That may be the case, but if you have any children, they might be wise to think twice before considering routine radiology as a career.

Killer robots feature in many sci-fi films (e.g. the Terminator series). There is even a scary video warning of Slaughterbots that could target individuals with unacceptable political views.[41] AI is now becoming increasingly used on the real battlefield, backed by enormous amounts of money spent on research and development.[42] It has led to some unexpected findings. Even a decade ago there were instances of soldiers who were almost inconsolable at the thought that damaged robots with which (or should we say with whom?) they had worked on the battlefield might not be repairable.[43] Peter Singer recounts how 'One EOD [Explosive Ordnance Disposal] soldier brought in a robot for repairs with tears in his eyes and asked the repair shop if it could put "Scooby-Doo" back together. Despite being assured that he would get a new robot, the soldier remained inconsolable. He only wanted Scooby-Doo.'[44] Another soldier ran 50 m under machine gun fire to rescue a robot that had been knocked out by enemy fire.[45] On the battlefield, robots have been promoted, given Purple Heart awards, and received a military funeral.

What do the huge resources that are being invested in autonomous lethal robots say about our moral hope for peace? The history of technology shows how technologies change us and make it more difficult to resist the use of new technologies once their deployment become feasible—technologies are not like the individual dishes on a sushi train that we freely choose to accept or let pass by.[46]

Wisdom and values in the use of machine learning

Algorithmic justice is not new. In the Hebrew Deuteronomic law, if a man had sex with a woman who was engaged to be married to another man, then this was an unconditionally capital offence for the man. For the woman it depended on the circumstances.[47] If it occurred in a city, then she would be regarded as culpable, on the grounds that she should have screamed for help. But if it occurred in the open country, then she was to be presumed innocent, since however loudly she might have cried out for help there would have been no one to hear her. This is a kind of algorithmic justice: IF in city THEN woman guilty ELSE woman not guilty.

There are already plenty of areas where decisions are made algorithmically. Accountants love it when governments introduce ever more complex tax regulations, since they can then charge ever larger fees to their clients for minimizing their tax bills. This is a skill that is ripe for machine learning to exceed human intelligence, as is the detection of tax fraud where India seems to be taking the lead. Nevertheless, once the figures have been submitted to the authority, the tax one owes is determined algorithmically. An individual citizen declares their income (or their employer declares it for them); account is taken of allowances for which they qualify, and their tax bill is then calculated through an algorithm which at best is applied efficiently and impartially. There is much to be said for such algorithms; in regimes where the tax collector has too much human discretion, corruption is usually not far behind.

Computer algorithms work well for things like calculating tax, where it is a matter of applying the tax code, and for games like Go and chess, where the rules of the game are precise and there is no uncertainty in the position of the pieces. We have also seen that machine learning is rapidly becoming the equal of humans, or even surpassing us, for medical diagnoses. This marks a transition from dealing with certain data to dealing with uncertain data, and that is something that machine learning is very good at. You could even describe machine learning as the technology for managing uncertainty, with Bayesian optimization as the technology for making the best decisions with imperfect and incomplete data. In a world where it seems that increasingly important decisions are made on the basis of information which can be subject to vastly varying degrees of uncertainty, wisdom will be needed to discern when machines will do a better job than humans of managing the uncertainty, and when the uncertainty is so radical or of such a kind that the best approach is for humans to stand back and ask simply, 'What is going on here?'[48]

What about deciding how long a sentence should be for a prisoner found guilty (Figure 10.2)? Beyond uncertainty in the data, many decisions involve moral values which science is unable to supply. In work the two of us have undertaken with Dominic Burbidge, we point out that sentencing necessarily goes beyond the facts in that it requires judgements to be made and that entails wisdom.[49] We also recognize that judicial decisions require human involvement not only to be correct (insofar as that is possible in messy human situations) but also to be

Figure 10.2 Kathleen Gentry reacting to her sentencing of four years in prison, after being found guilty, with her husband, of charges of involuntary manslaughter, felony child endangerment, and conspiracy for the 1996 death of their disabled daughter. The two men shown are her attorneys.

acceptable. Justice needs not only to be done and to be seen to be done; in an open society it needs also to meet with widespread acceptance.

A utilitarian approach to sentencing would have as its fundamental premise that sentence length should be such that the aggregate of human happiness is maximized. If (as is thankfully the case) most people never commit rape and are confident that they won't, the sentence for rape that might maximize human happiness might be life sentences for rapists, since the unhappiness of the convicted rapists would be likely to be more than outweighed by the happiness of the rest of the population whose fear of rape is thereby ameliorated. The death penalty might be even more effective, by eliminating the possibility of escape and saving money. But utilitarianism is a blunt ethical instrument and can be unfair to minorities, such as rapists. Taking away a person's freedom should be based on considerations of justice.

That is why boundaries are set by sentencing guidelines. You can try these for yourself. The Sentencing Council in England and Wales has a website that offers eight real-life court cases, in which the reader can learn the facts of each case and the aggravating and mitigating factors, and with reference to the relevant sentencing guidelines decide what they think the sentence should be.[50] You can then compare your judgment with the sentence handed out by the judge in the real case.

In the case of rape in the UK, the guidelines have as their starting point:

- Single offence of rape by single offender: 5 years custody if victim 16 [years of age] or over; 8 years custody if victim 13 or over but under 16; 10 years custody if victim under 13.
- Rape accompanied by aggravating factor (e.g. abduction, offender aware they have a sexually transmitted infection, abuse of trust): 8 years custody if victim 16 or over; 10 years custody if victim aged 13 or over but under 16; 13 years custody if victim under 13.
- Repeated rape of same victim by single offender or rape involving multiple victims: 15 years custody.[51]

However, these are only the starting points. In all cases the maximum penalty is life imprisonment. Imagine that we decide that the sentence length between these starting points and life imprisonment should be determined not by a human judge but by a machine. We would need to provide the machine with a criterion. If our criterion were 'minimize the chances of re-offending', it is likely the machine would simply sentence all convicted rapists to life imprisonment. The same outcome might result from a criterion of 'minimize emotional hurt to the victim'. But suppose we have a more complicated criterion such as 'overall, do not allow the prison population at any time to exceed N, while minimizing the overall expected number of re-offences'. Now we have something that machine learning can get its teeth into. By exhaustively trawling (something machines are good at) historical data on re-offending rates as a function of sentence length, along with modelling future prison numbers, machine learning might indeed be able to come up with recommended sentences.

Might machine learning one day be able to recommend fairer sentences than human judges? Legal systems, in the UK and elsewhere, depend on interpreting the written law through case law, drawing on judgments that have been made in similar cases. Machine learning is rather good at that kind of thing. Braden Allenby and Daniel Sarewitz argue that we need to maintain a commitment to fundamental human values and embrace a new techno-human relationship and search for a new humility.[52] Techniques are also being developed for machines to learn human values. The machine does not have to have a comprehensive value system of its own; rather, it should be able accurately to learn the value system for the decisions which it is being asked to make.[53] Machines are now able to learn human values through inverse reinforcement learning, whereby machines learn an agent's values by observing its behaviour.

Imagine that you are trying to train an autonomous vehicle to drive safely. One way to do this is to let the car drive with you as an observer. It would be best to start in a practice area before moving into traffic. When the machine drives the car safely, you reward the algorithm. When the machine gets the car into danger you take over the controls, and penalize the algorithm. This is reinforcement learning, and it is not very different from how humans are taught to drive by an instructor. Imitation learning works the other way around. You drive the car to the highest standards you know. The machine makes moment by moment predictions of what you will do next. When the machine predicts correctly, you reward it, and when it predicts wrongly you penalize it. When the machine is able to predict with sufficiently high accuracy what the human would do, you give it its driving licence. You can see how this could be applied to sentencing decisions. The machine would have superhuman capacity to learn all the relevant law and study all the relevant cases.

Inverse reinforcement learning takes this a notch higher.[54] The machine now seeks not simply to *imitate* the agent, but to learn its *values* by observing its behaviour. If this sounds impossible, think how Amazon might do this simply from our shopping habits. If it finds that a person regularly buys flowers and chocolates for delivery to their home address on the same day each year, it may deduce something about the relationships in the household. It may even be able to tie that, from publicly available records, into the birth and marriage dates of another person living at the same address. Combine that with all the other data available about our *behaviours*, and the machine may begin to learn about our *values*.

The output from machine learning is limited by the quality of the data that are inputted. Suppose, for example, that there are existing biases in sentence duration given to offenders by virtue of their ethnicity or gender. Unless explicit care is taken to account for such biases, they are likely simply to be reinforced.[55] This is not only a problem for machines, which is why humans need to undergo hidden bias training. As we write this chapter following the death of George Floyd in the USA and the removal of slave-trader statues in the UK, there are no grounds for complacency about human bias. It is an open question whether hidden bias in machines will prove harder or easier to correct than hidden bias in humans.

Following the UK Coroners and Justice Act 2009,[56] detailed criteria are available to apply to the facts of each case in which there is a verdict of guilty.[57] A judge is even required to consider mitigating factors, including the extent to which there is 'genuine remorse'.[58] Could a machine show mercy in sentencing? It has been suggested that a form of 'artificial reason' might be needed for the exercise of mercy within a context of justice, although this was pre-AI artificial reason as an attribute of law, as opposed to the natural reason exercised by ordinary folk.[59] Mercy is one possible ground for leniency, focussed on the benefit to the miscreant; there can be others focussed on promoting social peace and democracy, which in part motivated the Truth and Reconciliation Commission following the apartheid era in South Africa. Different judges can show more or less mercy when faced with the same set of facts and circumstances. If society accepts such variability from human judges, what kind of discretion would be acceptable from machines? We are not there yet, but the day may come when machines might be able to learn about values in the judicial system (the kind of values of justice which we hold dear) from the decisions made by judges and the published accounts of the cases and of the laws which underpin them, for many more cases than a human would ever be able to read up.

At this point the reader may be sucking air through their intellectual teeth. The source of such unease may be the use to which machine learning is put for commercial and political ends.[60] If you think that machine learning is not already being applied to you, you are probably mistaken. Almost every time you do an online search or use social media, the big data companies are harvesting your data exhaust for their own ends. Even if your phone calls and emails are secure, they still generate metadata. European legislation is better than most, but it is limited in what it protects. Targeted persuasion predates AI, as Othello's Iago knew, but machine learning has brought it to an unprecedented level of industrialization, with some of the best minds in the world paid some of the highest salaries in the world to maximize the user's screen time and the personalization of commercial and political influence.[61]

An even deeper source of unease may be because of an underlying conviction that decisions that affect the bodies and the freedom of humans ought to be taken by humans. But which would you rather be diagnosed by? An established human radiologist, or a machine with demonstrated superior performance? Consider this step by step. Would you want to be diagnosed by a machine that knew less than your doctor? Answer: 'No!' Well then, would you want to be diagnosed by a doctor who knew less than the machine? That's more difficult. Perhaps the question needs to be changed. Would you prefer to be treated by a doctor without machine learning or by a doctor making wise use of machine learning?

If we want humans to be involved in decisions involving our health, how much more in decisions involving our freedom? But are humans completely reliable and consistent? A peer-reviewed study suggested that the probability of a favourable parole decision depended on whether the judges had had their lunch,[62] though the study does not seem entirely to stand up to scrutiny.[63] The very fact that appeals are sometimes successful provides empirical evidence that law, like any other human endeavour, involves uncertainty and fallibility. Would you want a machine that is less consistent than a judge to pass sentence? See the sequence of questions above about a doctor. If machine learning is to contribute as it should to human flourishing, humans will need to combine the best of scientific insight with the best of spiritual wisdom.

Gene technologies

There are dangers in seeing genes as key to all of biology.[64] Without the proteins and other components of the cells in which they normally exist, in turn embedded in a functioning organism, genes can do nothing.[65] Nevertheless, genes have one crucial property which other components of life lack: they persist virtually unchanged for many, many generations. Changes to the genetic material (usually DNA but in some viruses RNA) can have almost permanent consequences. As with AI, issues of safety and values are to the fore in the various gene technologies. In addition, there can be a profound concern that we are hubristically giving ourselves powers beyond our remit—'playing God', as it is often put.

Conventional genetic engineering

In recent decades it has become possible for scientists to alter the genetic material of organisms. Most methods of genetic engineering, and certainly those of greatest concern both to the general public and to members of pressure groups opposed to genetic engineering, involve scientists moving genes between species, which are often completely unrelated.[66] For example, the gene in humans responsible for the production of the hormone insulin has been moved into bacteria or yeast. As a result, genetically engineered human insulin is now widely used in the treatment of people suffering from juvenile-onset diabetes (Figure 10.3).

Moving genes from humans to bacteria or yeast, as is done to make genetically engineered human insulin, seems unnatural. This will not worry some. On one understanding, where we are nothing but the materials of which we are made and our choices are merely the products of our neurons mindlessly firing, everything is natural. In that sense all toolmaking is natural, and a teenager using a smartphone to text her friends is no more artificial than a Caledonian crow using a stick to extract insects from tree bark. It is only as responsible agents that humans can engage in interventions that can be described as artificial, and on that basis evaluated for their moral contribution to human flourishing.

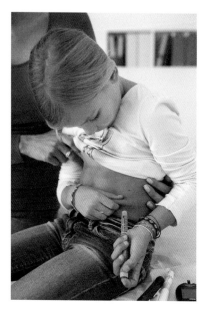

Figure 10.3 Most of the insulin injected by people with diabetes is genetically engineered so that it is more similar to human insulin than the insulin obtained from cows or pigs.

Although hybridization between closely related species and so-called 'horizontal gene transfer' does occur in nature—meaning that the genetic boundaries between species are not as absolute as is sometimes supposed—genetic engineering breaches species boundaries in ways that nature does not. How concerned should we be at this? Does it matter that plant crops contain bacterial or animal genes if the result is that their yields are greater? Does it matter that certain bacteria confined to fermenters in pharmaceutical factories contain human genes if the result is that life-saving and health-restoring medicines are produced? Does it matter that pigs are being genetically engineered with human genes in the hope that their internal organs may be used for human transplants?[67]

As each of us develops, the boundaries between species help us to organize our understanding of the natural world. Children learn from their infancy about living things in their immediate environment. In particular, they learn about animals, learning to recognize different types of animals and what their basic names are.[68] The concepts 'animal' and 'plant' are fundamental categories for children to organize their perceptions of the world in which they live.[69] Animals form a significant part of the world around most children as wildlife, pets, or zoomorphic toys. Children very early learn to identify and classify animals, and names for familiar animals form a large part, sometimes the largest part, of the vocabulary of young children.[70]

Boundaries manifest themselves differently in different societies. We distinguish between food and non-food, foreigner and non-foreigner, male and female, appropriate marriage partners and inappropriate marriage partners. One culture thinks dogs are delicious to eat, another eyeballs, another haggis, another hákarl (the rotting carcass of Greenland or basking shark), and some cultures prefer first cousin marriages and others forbid it. Establishing personal boundaries is essential for human flourishing.[71] What theological significance do boundaries have?

In the Jewish scriptures, one of the key themes is the formation and maintenance of the state of Israel. The boundary in Judaism between Jews and non-Jews at times became so rigid that some of the Jewish prophets encouraged their listeners to look forward to a time when Israel's God would be worshipped throughout the whole world. For Christians, this promised time was ushered in by the life, work, death, resurrection, and ascension of Christ. It was very hard for the early Jewish Christians to know how to handle the boundaries with which they had grown up. What about the food laws, which distinguished what was acceptable and unacceptable to eat? What about circumcision, the definitive mark of a male who was part of God's chosen people? And what about specific aspects of the Jewish law such as how to keep the Sabbath as a special day of spiritual regeneration? These and other topics were the subject of a landmark conference in Jerusalem.[72] The outcome combined a recognition of individual sensibilities with a breakthrough in conceptual understanding. As Paul was to put it, in his manifesto for universal human rights: 'There is neither Jew nor Gentile, neither slave nor free, nor is there male and female, for you are all one in Christ Jesus.'[73]

For a particular moral issue raised as a result of the movement of genes between species in genetic engineering, consider the question: is it wrong to eat animals that have been genetically engineered so that they contain human genes? Imutran, one of the companies that once actively engaged in xenotransplantation research (which meant that pigs were genetically engineered for medical reasons with a small number of human genes), argued that 'This involves changing only 0.001% of the genetic make-up of the pig.'[74] Furthermore, if it were absolutely wrong to consume human DNA, then all babies should be forbidden to breastfeed, since that involves consuming large amounts of their mothers' DNA. However, just because a baby does certain things with its mother doesn't make it right for the rest of us to do those same things with her.

If we were to refrain from any foods that share DNA with humans we should starve. We share about 90 per cent of our DNA with cats, 85 per cent with mice, 80 per cent with cows, and 60 per cent with insects.[75] We even share 60 per cent with bananas, so don't eat bananas, indeed plants in general, if you want to avoid large amounts of DNA that are also found in humans! Those who draw parallels between the genetic engineering of foods that we eat using human genes and cannibalism might argue that it is the very presence of human genetic material that matters, not the percentage. Being unfaithful to one's spouse on 0.5 per cent of nights is not ten times less bad than being unfaithful on 5 per cent of nights. It may be that rather as ethical debates about embryonic stem cells have been modulated by the availability of adult stem cells, so ethical issues about human using human genes in food will be overtaken by developments in gene editing and gene synthesis.

Gene therapy

Not all genetic engineering entails moving genes between species. Gene therapy is a set of techniques, still mostly experimental, that alter human genes to prevent or treat diseases. It does not require genes from other species to be introduced into the human genome; rather, it is all about repairing damaged genes so that their normal function is restored. An important distinction is between somatic gene therapy, in which alterations are made to the 'ordinary' cells in the body, such as the cells in the heart, the digestive system, or the nervous system, and germ-line gene therapy, in which alterations are made to the specialized cells in the ovaries and testes that give rise to our gametes (eggs and sperm). At present, germ-line gene therapy is not permitted in most countries, even for research purposes. That may slowly change if the technique is shown to be safe enough and sufficiently well-understood to be used to save future generations from inherited diseases.

Figure 10.4 Jesse Gelsinger (18 June 1981–17 September 1999), who died while enrolled in a gene therapy trial.

As with a number of gene technologies, the early history of gene therapy was characterized more by hype than instances of successful treatments. In 1999 Jesse Gelsinger joined a clinical trial of a somatic gene therapy aimed at curing ornithine transcarbamylase deficiency, a rare genetic disease of the liver (Figure 10.4). This condition is usually fatal at birth, but Gelsinger had a mild form and was able to survive, thanks to a special diet and medication. He died four days into the trial having suffered a massive immune response triggered by the use of a viral vector to transport the healthy version of the gene into his cells, leading to multiple organ failure and brain death.[76] The ensuing investigation by the Food and Drug Administration identified several rules of conduct which had been broken.

In 2020 the Nobel Prize in Chemistry—for achievements that have conferred the greatest benefit to humankind—was awarded to Emmanuelle Charpentier and Jennifer Doudna 'for the development of a method for genome editing'.[77] This was the first time a Nobel Prize had been awarded to an all-female team. The CRISPR/Cas9 genetic scissors can be used to change DNA with extremely high precision. The

Figure 10.5 Christian Guardino and his mother, Beth, talk about his life before and after gene therapy for his hereditary blindness.

number of successful treatments of patients using gene therapy is steadily growing (Figure 10.5). Somatic gene therapy is now being used to treat patients with a number of inherited conditions, including severe combined immune deficiency, degenerative blindness, various blood diseases, and various metabolic disorders.[78] At present often the only 'treatment' for some genetic disorders is, if they can be detected before birth, to offer the mother the possibility of a termination of pregnancy. Gene therapy (somatic or germ-line) offers an alternative that many will find ethically preferable.

The most frequent response by religious writers to the issue of gene therapy in humans has been one of caution or hesitancy.[79] A more positive response is provided by Ron Cole-Turner, who has explored the implications of a distinction between humans as co-creators with God—a concept which, he feels, contains a number of difficulties, including hubris—and humans as participants, through gene therapy, in redemption.[80] Here, redemption is being used in the sense of 'restoration'. If gene therapy can help to overcome genetic defects caused by harmful mutations, then it can restore humans to a fuller, richer existence. Humans might therefore have a theological responsibility, even a duty, to use gene therapy to root out imperfections in the natural world, including those found in humans. As a tool with the potential to eliminate harmful genetic mutations, gene therapy can reduce suffering and thereby contribute to restoring creation to its full glory.[81]

The concept of 'playing God' is often used pejoratively as though it were self-evidently a *Bad Thing*. But like many slogans it risks substituting knee-jerk reaction for careful thought and analysis. The Hebrew concept of צֶלֶם אֱלֹהִים, often translated as 'image of God', or sometimes in Latin as *Imago Dei*, carries a strong connotation of acting on God's behalf, as God's vicegerent.[82] In the right sense, 'playing God' is exactly what humans are called to do. This applies as much to life-saving interventions as it does to every other sphere of human endeavour. What is at issue, for those with a religious faith, is not whether humans should take hard decisions and implement them, but whether in making those choices they promote human flourishing in the way that God intends.

Genetic enhancement

Transhumanism is the notion that humanity can grow beyond its present limitations through the applications of science and technology. The term seems first to have been used by the biologist Julian Huxley (the brother of Aldous Huxley, author of *Brave New World*) in 1927:

> The human species can, if it wishes, transcend itself—not just sporadically, an individual here in one way, an individual there in another way—but in its entirety, as humanity. We need a name for this new belief. Perhaps transhumanism will serve: man remaining man, but transcending himself, by realizing new possibilities of and for his human nature.[83]

It may be widely accepted that interventions can be made, whether genetic or not, to help restore people to health, so long as these are safe and those being treated consent. But what about interventions intended to enhance human traits or capacities? John Habgood, Archbishop of York from 1983 to 1995, was a formidable public intellectual, with an academic career in pharmacology before being ordained.[84] Habgood cautioned against attempts at using genetic engineering to improve people, concluding with six succinct rules:

> First, human beings are more than their genes. Genes are only a set of instructions. We are more than a set of instructions. Second rule: remember the valuable diversity of human nature. Third rule: look for justice in the

dealings of human beings with one another and for fairness in the use of resources. Fourth rule: respect privacy and autonomy. Fifth rule: accept the presumption that diseases should be cured when it is possible to do so. And sixth rule: be very suspicious about improving human nature; and be even more suspicious of those who think they know what improvements ought to be made.[85]

It is not only religious leaders who have warned of the dangers of presuming to improve human nature or enhance human capabilities. Jonathan Glover quotes the philosopher John Mackie who once argued, against Glover's optimism about germ-line therapy, that 'if the Victorians had been able to use genetic engineering, they would have aimed to make us more pious and patriotic.'[86]

Genetic enhancement can be thought of as an instance of transhumanism. The theologian Michael Burdett points out that whereas transhumanists generally treat death as a disease to be overcome, religions view death differently.[87] For all that death, in the Christian tradition, is seen as an enemy, it is overcome not in endless transplants, gene editing, or cryopreservation, which can be seen as merely the promises of a secularized eschatology, but in the resurrection. This puts a different perspective on genetic enhancement. Instead of asking, 'How can we use genetic technology to avoid death?', which as far as we know is an unachievable goal, the question becomes 'How can we use genetic technology to bring true human flourishing?' The difference between the two is as wide as the world.

Gene synthesis

If gene therapy is rather like making corrections to the proofs of a publication provided by a publisher, and gene editing is rather like amending a draft of a paper written by a fellow author, gene synthesis is more like sitting at your computer with a blank document and writing a completely new piece. You can literally input the desired sequence of bases, and the machine will produce the corresponding DNA. Gene synthesis gives us the possibility to construct gene sequences that have never been seen before in nature.[88] It is even possible to use building blocks that go beyond the four DNA bases of adenine, thymine, guanine, and cytosine that suffice for all of life as we know it.[89]

Andrew is on the scientific advisory board of a company which achieves gene synthesis through a combination of thermal purification and repeated doubling of length.[90] The key is precise temperature control of a large number of pixels on a silicon chip. On each pixel a strand is grown by thermally removing a protective cap to expose it to the next base to be attached. After the strand reaches a suitable length, a matching strand is attached to it. This double strand is then heated to a temperature at which if it is perfect it will survive, but if there is any mismatch the two strands will separate, thereby eliminating faults. Double strands that have thus been proven to be error-free can then be attached end to end, to produce a new double strand of twice the length. Like the legendary Chinese peasant who asked the emperor for one grain of rice on the first square of a chess board, two on the second, and so on, the repeated adding of sections each twice the length of its predecessor leads exponentially to long sequences of DNA, which have at each stage been checked to be error-free. This opens the way to applications in healthcare such as new antibodies and vaccines, industrial biotechnology for products currently produced from oil, new biodegradable materials, agricultural crops and fertilizers offering greater yield and less cost to the environment, and long-lasting molecularly dense data storage.

Even more than artificial general intelligence, gene synthesis of anything as complex as a human will remain science fiction for the foreseeable future. Nevertheless, even contemplating how we would wish to use technologies to change the characteristics of humans if we could can help us to focus on what constitutes human flourishing. Consider a thought experiment in which musical talent could be enhanced genetically over a population. The thought experiment starts with three premises:

1. Communal music making can contribute to human flourishing, whether it be the West-Eastern Divan Orchestra founded by Edward Said and Daniel Barenboim,[91] El Sistema music programme founded by José Antonio Abreu,[92] or local choirs that flourish the world over.
2. Where children are given the opportunity to make music, it is remarkable how many of them join in.
3. There seems to be a hereditary component of musical ability—musical parents beget musical children.

No doubt much of the hereditary aspect is environmental, but let us suppose that some of it is genetic. And let us further suppose (for the sake of this fanciful thought experiment) that one day enough is learned about its polygenetic basis to be able safely to offer genetic enhancement to be able to improve the musical aptitude of a population. Would that be such a bad thing to do? If so, why? Is it at least conceivable that greater genetically based aptitude for music could actually contribute to human flourishing?

You can see where this is going. If it could be beneficial to enhance musical ability, what about other abilities? What about in areas that are often competitive, such as sport? Should pushy parents be allowed to enhance the potential athletic prowess of their offspring genetically? Racehorse owners have long been doing that by selective breeding and an increasing amount is known about the genetics of success in thoroughbreds.[93] What about other spheres of human endeavour that are competitive, such as gaining admission to elite universities? Would parents who could afford it be willing to pay large sums to improve their children's chances? You bet they would.

It is time to get real again. These flights of imaginitis can give us a stop-and-think point to ask what really matters in human flourishing. If we could genetically enhance the capacities of future generations, what capacities would we want to enhance? Would it be musical potential, athletic ability, physical appearance, IQ? All these are things that people strive after. Or would it be the capacity for whatever is true, whatever is noble, whatever is right, whatever is pure, whatever is lovely, whatever is admirable, whatever is excellent or praiseworthy?[94] Which, if any, would you choose?

COVID-19

We are writing this in the second UK lockdown for the COVID-19 pandemic. COVID-19 has implications for human flourishing and for the place of scientific insight and spiritual wisdom in addressing the disease. While COVID-19 is not a technology—and unlikely to have been caused by any technology (the 'escaped from a lab' thesis[95]), though its spread was hugely facilitated by modern transport technologies— tackling it will require the best use of present and future technologies. At the time that we write, it is too soon to know whether the world-wide disruption that COVID-19 has caused will be with us for years to come or will fade more quickly.

COVID-19 is sometimes compared with the 1918–19 influenza pandemic.[96] It has been estimated that about 500 million people became infected with that strain of the influenza virus (one-third of the then world's population) and about 50 million people died (a mortality rate of about 10 per cent), more than the number killed in the ghastly war over the preceding four years. Like COVID-19, the disease was another example of a zoonosis (a disease transmitted to humans from non-human animals), being caused by an H1N1 virus with genes of avian origin. Unlike COVID-19, where the risk of death increases with age, mortality from the 1918–19 influenza showed three peaks: in people aged less than 5 years old, 20–40 years old, and 65 years and older, for reasons which are not understood.[97]

Attempts to tackle COVID-19 can be compared with the historical attempts to tackle the 1918–19 influenza pandemic (Figure 10.6). Masks were worn, public gatherings banned, schools and businesses closed, good hygiene practices recommended, makeshift hospitals established, and desperate (unsuccessful) attempts made to manufacture a vaccine. In the end, it was herd immunity that caused the 1918–19 influenza to die out. If it is herd immunity that causes COVID-19 to die out, it will have cost the lives of many millions of people.

At the time of writing, it is difficult to be sure how successful vaccines against COVID-19 will be. A September 2020 poll of 2,730 US adults

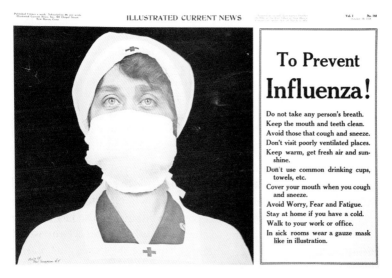

Figure 10.6 Similar strategies were used in the 1918–19 influenza pandemic and the early stages of the COVID-19 pandemic.

revealed that 61 per cent said they would likely get a COVID-19 vaccine.[98] Attitudes divided along party lines: 77 per cent of Democrat voters said they would, compared to 44 per cent of Republican voters. There is a similar polarization about wearing a face mask. People with objections to vaccinations are sometimes pilloried as ignorant or selfish. Objections to vaccination began almost as soon as the practice was introduced. Nineteenth-century objections included arguments that they didn't work and were unsafe[99] or that their compulsory introduction (e.g. the 1853 Compulsory Vaccination Act in the UK) violated personal liberties.[100] To this day, vaccination is rejected by some for much the same reasons. Good quality communication of the natural sciences and appreciation of psychology and sociology are needed, though it is unlikely that these alone will be sufficient. Polarized views are seldom due solely to an information deficit.[101] Other factors are almost invariably at play, often decisively.

The need for an interdisciplinary approach to tackling COVID-19, rather than one that claims only to be guided by 'the science' or one that largely ignore science, is impressively indicated by a 2013 article—written long before COVID-19 was known—about H5N1 avian influenza and Ebola:

> Zoonotic diseases currently pose both major health threats and complex scientific and policy challenges…[which] are best met by combining multiple models and modelling approaches that elucidate the various epidemiological, ecological and social processes at work. These models should not be understood as neutral science informing policy in a linear manner, but as having social and political lives: social, cultural and political norms and values that shape their development and which they carry and project.…Addressing the complex, uncertain dynamics of zoonotic disease requires such social and political lives to be made explicit in approaches that aim at triangulation rather than integration, and plural and conditional rather than singular forms of policy advice.[102]

To social and political we can add religious lives. In the first half of 2020, major religious festivals—Pessach, Easter, Vaisakhi, Navratri, and Ramadan—were celebrated under lockdown. Some public health policy makers were sceptical that physical distancing, the key strategy for most countries struggling to contain COVID-19, could be achieved during religious festival gatherings. In South Korea, a church congregation is thought to have been the origin of thousands of first cases in the country. In Malaysia, a large Islamic gathering of the Tablighi-Jamaat of up to 14,000 delegates in Kuala Lumpur was widely considered the cause of the second wave in the trajectory of the epidemic in the country. Attendees returning to Brunei, Indonesia, and Cambodia from this gathering later tested positive for COVID-19. And yet, most religious adherents the world over have complied.[103]

At their best, religions can be enablers of public health, and religious organizations can be important partners, especially in less secular societies. A pluralist society necessarily comprises different worldviews. Atheists may find it incomprehensible that a person disregards substantial risk of infection in order to attend an 'in person' service. Religious communities may believe that some public health initiatives are directly opposed to their freedom to worship and can feel that they, rather than the virus, are under attack. But while services may not be seen as 'essential' by some, others will place a high value on them, all the more so in the light of the evidence that religion can serve as an important coping mechanism and a source of emotional support.[104]

In ways that none of us would have chosen, COVID-19 has forced the world to rethink its values. We have seen extraordinary acts of selflessness, from the heroic commitment of health care workers to countless acts of kindness by one neighbour to another. We have seen selfishness and carelessness, as people needlessly socialize in defiance of national or local regulations. We have even seen a major election polarized according to whether to wear a face mask. And we have seen serious debates evaluating the tension between containing the spread of infection and mitigating the consequent economic damage.

In promoting human flourishing through the best of scientific insight and spiritual wisdom, we have sought to provide a basis for thinking about what really matters. When the next existential threat arises, and when new technological possibilities emerge, we hope that what we have provided will enable good priorities to be set, and good choices to be made. In the concluding chapter, we set out an even better way.

Notes

1   Briggs, G. A. D. (1992) *The Science of New Materials*, Oxford: Blackwell, p. 1.

2   Eddington, A. S. (1928) *The Nature of the Physical World*, Cambridge, UK: Cambridge University Press, p. 74.

3   Harmand, S., Lewis, J., Feibel, C. et al. (2015) 3.3-million-year-old stone tools from Lomekwi 3, West Turkana, Kenya, *Nature* **521**, 310–15.

4 James, S. R. (1989) Hominid use of fire in the Lower and Middle Pleistocene: A review of the evidence, *Current Anthropology* **31**(1), 1–26.

5 Zeder, M. (2011) The origins of agriculture in the Near East, *Current Anthropology* **52**(S4): S221–35.

6 Anthony, D. A. (2007) *The Horse, the Wheel, and Language: How Bronze-Age Riders from the Eurasian Steppes Shaped the Modern World*, Princeton, NJ: Princeton University Press.

7 Humphreys, C. J. (1992) Can there be a materials policy in the UK?, in G. A. D. Briggs (Ed.), *The Science of New Materials*, Oxford: Blackwell, pp. 177–8.

8 Van Lawick-Goodall, J. (1971) *In the Shadow of Man*, Glasgow: Collins.

9 Jelbert, S. A. (2014) Using the Aesop's Fable paradigm to investigate causal understanding of water displacement by New Caledonian crows, *PLOS ONE* **9**, e92895.

10 Jelbert, S. A. (2019) New Caledonian crows infer the weight of objects from observing their movements in a breeze, *Proceedings of the Royal Society B* **286**, 20182332.

11 Agricola, G. (1556) *De Re Metallica*, trans. H. C. Hoover and L. H. Hoover (1912) London: The Mining Magazine, p. 18.

12 Browne, J. (2019) *Make, Think, Imagine: Engineering the Future of Civilisation*, London: Bloomsbury, p. 4.

13 Fenn, J. and Blosch, M. (2018) Understanding Gartner's Hype Cycles, Gartner Research. Available at https://www.gartner.com/en/documents/3887767.

14 Vallor, S. (2016) *Technology and the Virtues: A Philosophical Guide to a Future Worth Wanting*, Oxford: Oxford University Press, p. 2.

15 Susskind, D. (2020) *A World without Work: Technology, Automation and how we should respond*, London: Allen Lane, p. 165.

16 Elliott, S. W. (2017) Computers and the future of skill demand, *OECD Educational Research and Innovation*, p. 17.

17 McCarthy, J., Minsky, M. L., Rochester, N. and Shannon, C. E. (1955) A Proposal for the Dartmouth Summer Research Project on Artificial Intelligence, p. 1. Available at http://raysolomonoff.com/dartmouth/boxa/dart564props.pdf.

18 Turing, A. M. (1950) Computing machinery and intelligence, *Mind* **59**, 433–60.

19 Christian, B. (2012) *The Most Human Human: What Artificial Intelligence Teaches Us about Being Alive*, London: Penguin, p. 190.

20 Searle, J. R. (1980) Minds, brains, and programs, *Behavioral and Brain Sciences* **3**(3), 417–57, p. 417.

21 Bostrom, N. (2014) *Superintelligence: Paths, Dangers, Strategies*, Oxford: Oxford University Press.

22 Shanahan, M. (2105) *The Technological Singularity*, Cambridge, MA: MIT Press.

23 Tegmark, M. (2018) *Life 3.0: Being Human in the Age of Artificial Intelligence*, London: Penguin.

24 McEwan, I. (2019) *Machines Like Me*, London: Jonathan Cape.

25 Wang, P (2019) On defining artificial intelligence, *Journal of Artificial General Intelligence* **10**, 1–37.

26 Kushmaro, P. (2018) 5 ways industrial AI is revolutionizing manufacturing, CIO, 27 September. Available at https://www.cio.com/article/3309058/5-ways-industrial-ai-is-revolutionizing-manufacturing.html.

27 Joshi, N. (2019) How AI can transform the transportation industry, *Forbes*, 26 July. Available at https://www.forbes.com/sites/cognitiveworld/2019/07/26/how-ai-can-transform-the-transportation-industry/#20cd3f014964.

28 Briggs, A. (2017) Why artificial intelligence will enable new scientific discoveries, Graphcore, 27 November. Available at https://www.graphcore.ai/posts/why-artificial-intelligence-will-allow-us-to-make-new-scientific-discoveries. See, for example, Lennon, D. T., Moon, H., Camenzind, L. C., Yu, L., Zumbühl, D. M., Briggs, G. A. D. et al. (2019) Efficiently measuring a quantum device using machine learning, *npj Quantum Information* **5**, 79.

29 Waterson, J. (2020) Microsoft sacks journalists to replace them with robots, *The Guardian*, 30 May. Available at https://www.theguardian.com/technology/2020/may/30/microsoft-sacks-journalists-to-replace-them-with-robots.

30 Christakis, N. A. (2019) *Blueprint: The Evolutionary Origins of a Good Society*, New York: Little, Brown Spark.

31 Reiss, M. J. (2021) The use of AI in education: Practicalities and ethical considerations, *London Review of Education* **19**(1), 5, 1–14.

32 Silver, D., Schrittwieser, J., Simonyan, K., Antonoglou, I., Huang, A., Guez, A. et al. (2017) Mastering the game of Go without human knowledge, *Nature* **550**, 354–9.

33 Anon (2015) Google's AlphaGo gets 'divine' Go ranking, *The Straits Times*, 15 March. Available at https://www.straitstimes.com/asia/east-asia/googles-alphago-gets-divine-go-ranking.

34 Silver, D. Huang, A., Maddison, C. J., Guez, A., Sifre, L., van den Driessche, G. et al. (2016) Mastering the game of Go with deep neural networks and tree search, *Nature* **529**, 484–9.

35 Silver, D., Hubert, T., Schrittwieser, J., Antonoglou, I., Lai, M., Guez, A. et al. (2018) A general reinforcement learning algorithm that masters chess, shogi, and Go through self-play, *Science* **362**, 1140–4.

36 Senior, A. W., Evans, R., Jumper, J., Kirkpatrick, J., Sifre, L., Green, T. et al. (2020) Improved protein structure prediction using potentials from deep learning, *Nature* **577**, 706–10.

37 Rodriguez-Ruiz, A., Lång, K., Gubern-Merida, A., Broeders, M., Gennaro, G., Clauser, P. et al. (2019) Stand-alone artificial intelligence for breast cancer detection in mammography: Comparison with 101 radiologists, *Journal of the National Cancer Institute* **111**, 916–22.

38 Padhy, S. K., Takkar, B., Chawla, R. and Kumar, A. (2019) Artificial intelligence in diabetic retinopathy: A natural step to the future, *Indian Journal of Ophthalmology* **67**(7), 1004–9.

39 McKinney, S. M., Sieniek, M., Godbole, V., Godwin, J., Antropova, N., Ashrafian, H. et al. (2020) International evaluation of an AI system for breast cancer screening, *Nature* **577**, 89–94.

40 The Royal College of Radiologists (2020) Thinking about a career in clinical radiology? Available at https://www.rcr.ac.uk/clinical-radiology/careers-recruitment/thinking-about-career-clinical-radiology.

41 Stop Autonomous Weapons (2017) Slaughterbots [video], YouTube, uploaded 12 November, https://www.youtube.com/watch?v=9CO6M2HsoIA.

42 Reding, D. F. and Eaton, J. (2020) *Science & Technology Trends 2020–2040: Exploring the S&T Edge*, NATO Science & Technology Organization. Available at https://www.sto.nato.int/pages/tech-trends.aspx, Appendix C, pp. 59–68.

43 Reiss, M. J. (2020) Robots as persons? Implications for moral education, *Journal of Moral Education* **50**(1), 68–76.

44 Hsu, J. (2009) Real soldiers love their robot brethren, *Live Science*, 21 May. Available at https://www.livescience.com/5432-real-soldiers-love-robot-brethren.html.

45 Singer, P. W. (2009) *Wired for War: The Robotics Revolution and Conflict in the 21st Century*, New York: Penguin.

46 Vallor, S. (2016) *Technology and the Virtues: A Philosophical Guide to a Future Worth Wanting*, Oxford: Oxford University Press.

47 Deuteronomy 22: 23–7.

48 Kay, J. and King, M. (2020) *Radical Uncertainty: Decision-Making for an Uncertain Future*, London: Bridge Street Press.

49 Burbidge, D., Briggs, A. and Reiss, M. J. (2020) *Citizenship in a Networked Age: An Agenda for Rebuilding Our Civic Ideals*, Oxford: University of Oxford. Available at https://citizenshipinanetworkedage.org/wp-content/uploads/2020/04/CiNA-Report-for-Web-with-Links.pdf.

50 Sentencing Council (2020) You be the Judge. Available at https://www.sentencingcouncil.org.uk/about-sentencing/you-be-the-judge/.

51 Crown Prosecution Service (2017) *Rape and Sexual Offences*, Chapter 19: Sentencing. Available at https://www.cps.gov.uk/legal-guidance/rape-and-sexual-offences-chapter-19-sentencing.

52 Allenby, B. R. and Sarewitz, D. (2011) *The Techno-human Condition*, Cambridge, MA: MIT Press.

53 Russell, S. (2019) *Human Compatible: Artificial Intelligence and the Problem of Control*, London: Penguin Random House.

54 Russell, S. and Norvig, P. (2020) *Artificial Intelligence: A Modern Approach*, 4th edn, London: Pearson; Ng, A. Y. and Russell, S. J. (2000) Algorithms for inverse reinforcement learning, *ICML '00*, 663–70.

55 There is now a growing literature on how such unwanted algorithmic bias can be mitigated, e.g. Bellamy, R. K. E., Dey, K., Hind, M. et al. (2019) AI fairness 360: An extensible toolkit for detecting, understanding, and mitigating unwanted algorithmic bias, *IBM Journal of Research and Development* **63**, 4:1–4:15.

56 Legislation.gov.uk (2009) Coroners and Justice Act 2009. Delivered by the National Archives. Available at http://www.legislation.gov.uk/ukpga/2009/25/contents.

57 Sentencing Council (2020) Sentencing Guidelines for use in Crown Court. Available at https://www.sentencingcouncil.org.uk/crown-court/.

58 Masslen, H. (2015) *Remorse, Penal Theory and Sentencing*, Oxford: Hart Publishing.

59 Tasioulas, J. (2003) Mercy, *Proceedings of the Aristotelian Society* **103**, 101–32.

60 Smith, B. and Browne, C. A. (2019) *Tools and Weapons: The Promise and the Peril of the Digital Age*, London: Hodder & Stoughton.

61 Zuboff, S. (2019) *The Age of Surveillance Capitalism: The Fight for a Human Future at the New Frontier of Power*, London: Profile Books. See also Howard, P. (2020) *Lie Machines: How to Save Democracy from Troll Armies, Deceitful Robots, Junk News Operations, and Political Operatives*, New Haven, CT: Yale University Press.

62 Danziger, S., Levav, J. and Avnaim-Pesso, L. (2011) Extraneous factors in judicial decisions, *Proceedings of the National Academy of Sciences* **108**, 6889–92.

63 Weinshall-Margel, K. and Shapard, J. (2011) Overlooked factors in the analysis of parole decisions, *Proceedings of the National Academy of Sciences* **108**, E833.

64 Reiss, M. J., Watts, F. and Wiseman, H. (Eds) (2020) *Rethinking Biology: Public Understandings*, Hackensack, NJ: World Scientific.

65 Alexander, D. R. (2020) *Are We Slaves to our Genes?*, Cambridge, UK: Cambridge University Press.

66 Reiss, M. J. (2003) Is it right to move genes between species? A theological perspective, in C. Deane-Drummond and B. Szersynski, with R. Grove-White (Eds), *Re-ordering Nature: Theology, Society and the New Genetics*, London: T. & T. Clark, pp. 138–50.

67 Reiss, M. J. (2000) The ethics of xenotransplantation, *Journal of Applied Philosophy* **17**, 253–62.

68 Rosch, E. and Mervis, C. B. (1975) Family resemblances: Studies in the internal structures of categories, *Cognitive Psychology* **7**, 573–605.

69 Keil, C. (1979) *Semantic and Conceptual Development: An Ontological Perspective*, Cambridge, MA: Harvard University Press.

70 Anglin, J. M. (1977) *Word, Object, and Conceptual Development*, New York: W. W. Norton.

71 Cloud, H. and Townsend, J. (2017) *Boundaries: When to Say Yes, How to Say No, to Take Control of Your Life*, Grand Rapids, MI: Zondervan.

72 Acts 15: 1–30.

73 Galatians 3: 28.

74 Novartis Imutran (1999) *Animal Welfare: Xenotransplantation—Helping to Solve the Global Organ Shortage*, Cambridge, UK: Imutran.

75 Deziel, C. (2018) Animals that share human DNA sequences, Sciencing, 20 July. Available at https://sciencing.com/animals-share-human-dna-sequences-8628167.html.

76 Stolberg, S. G. (1999) The biotech death of Jesse Gelsinger, *The New York Times Magazine*, 28 November, pp. 136–40, 149–50. Available at https://naberbiology.com/documents/BiotechDeathOfJesseGelsinger.pdf; Siddabld, B. (2001) Death but one unintended consequence of gene-therapy trial. *Canadian Medical Association Journal* **164**(11), 1612.

77 The Nobel Prize (2020) The Nobel Prize in Chemistry 2020. Available at https://www.nobelprize.org/prizes/chemistry/2020/summary/.

78 Genetic Science Learning Center (n.d.) Gene therapy successes. Available at https://learn.genetics.utah.edu/content/genetherapy/success/.

79 Reiss, M. J. (1999) What sort of people do we want? The ethics of changing people through genetic engineering, *Notre Dame Journal of Law, Ethics & Public Policy* **13**, 63–92.

80 Cole-Turner, R. (1993) *The New Genesis: Theology and the Genetic Revolution*, Louisville, KY: Westminster/John Knox Press.

81 Reiss, M. J. & Straughan, R. (1996) *Improving Nature? The Science and Ethics of Genetic Engineering*, Cambridge, UK: Cambridge University Press. Peters, T. (1997) *Playing God? Genetic Determinism and Human Freedom*, New York: Routledge.

82 Briggs, G. A. D., Halvorson, H. and Steane A. M. (2018) *It Keeps Me Seeking: The Invitation from Science, Philosophy and Religion*, Oxford: Oxford University Press.

83 Huxley, J. (1927) *Religion without Revelation*, London: E. Benn. Cited from Huxley, J. (1957) *Transhumanism*, London: Chatto & Windus, p. 17.

84 McGrath, A (2020) An undivided mind: John Habgood on science and religion, *Journal of Anglican Studies*, 1–13, doi:10.1017/S1740355320000224.

85 Habgood, J. (1995) Heslington Lecture given on 1 February at the University of New York.

86 Glover, J. (1984) *What Sort of People Should There Be?*, Harmondsworth, UK: Penguin, p. 149.

87 Burdett, M. S. (2015) *Eschatology and the Technological Future*, New York: Routledge.

88 Briggs, A. and Brears, T. (2019) Ethical considerations in an era of gene synthesis, *MedNous* **January**, 8–10.

89 Hoshika, H., Leal, N. A., Kim, M.-J., Kim, M.-S., Karalkar, N. B., Kim, H.-J. et al. (2019) Hachimoji DNA and RNA: A genetic system with eight building blocks, *Science* **22**, 884–7.

90 Evonetix Ltd, https://www.evonetix.com/.

91 West-Eastern Divan Orchestra, https://west-eastern-divan.org/.

92 Fundación del Estado para el Sistema Nacional de Orquestas Juveniles e Infantiles de Venezuela, https://www.fundacioncristinamasaveu.com/en/portfolio/national-system-of-youth-infant-orchestras-of-venezuela/.

93 Rooney, M. F., Hill, E. W., Kelly, V. P. and Porter, R. K. (2018) The 'speed gene' effect of *myostatin* arises in thoroughbred horses due to a promoter proximal SINE insertion, *PLoS ONE* **13**(10), e0205664.

94 Philippians 4: 8.

95 Cyranoski, D. (2020) The biggest mystery: What it will take to trace the coronavirus source, *Nature*, 5 June. Available at https://www.nature.com/articles/d41586-020-01541-z.

96 Reiss, M. J. (2020) Science education in the light of COVID-19: The contribution of history, philosophy and sociology of science, *Science & Education* **29**(4), 1079–92.

97 Centres for Disease Control and Prevention (2019) Influenza (Flu): 1918 Pandemic (H1N1 virus). Available at https://www.cdc.gov/flu/pandemic-resources/1918-pandemic-h1n1.html.

98 Largent, E. A., Persad, G., Sangenito, S., Glickman, A., Boyle, C. and Emanuel, E. J. (2020) US public attitudes toward COVID-19 vaccine mandates, *JAMA Network Open* **3**(12), e2033324.

99 Ernst, K. and Jacobs, E. T. (2012) Implications of philosophical and personal belief exemptions on re-emergence of vaccine-preventable disease: The role of spatial clustering in under-vaccination, *Human Vaccines & Immunotherapeutics* **8**(6), 838–41.

100 Durbach, N. (2000) They might as well brand us: Working class resistance to compulsory vaccination in Victorian England, *The Society for the Social History of Medicine* **13**(1), 45–62.

101 Haidt, J. (2012) *The Righteous Mind: Why Good People are Divided by Politics and Religion*, New York: Pantheon Books.

102 Leach, M. and Scoones, I. (2013) The social and political lives of zoonotic disease models: Narratives, science and policy, *Social Science and Medicine* **88**, 10–17, p. 10.

103 Barmania, S. and Reiss, M. J. (2020) How religion can aid public health messaging during a pandemic, *Nature India*, 20 May. Available at https://www.natureasia.com/en/nindia/article/10.1038/nindia.2020.87.

104 VanderWeele, T. J. (2017) On the promotion of human flourishing, *Proceedings of the National Academy of Sciences* **114**, 8148–56.

CONCLUSION
ACTIONABLE LOVE

Overview

An international speaker, writer, and activist identified three enduring virtues: faith, hope, and love. He reckoned that of the three the greatest is love.[1] Love needs to be given content in order to be actionable. If we had introduced the concept earlier in the book, there would have been insufficient background to appreciate what we meant. We trust that the stage is now set for the concluding drama of love as the fuel for flourishing.

Note

1 1 Corinthians 13: 1–13.

11

Human Flourishing Fuelled by Love

What has been missing so far? Love. How can we have written so much about human flourishing and said so little about love? In part that is because love has become such a malleable word that can be hard to use it in a specific context. More of that below. But in part it is because we have been saving the best until last. Human flourishing is love in action. Not only can flourishing not be complete without love, it cannot even begin without love.

The risk with a word that covers too much is that it may end up by conveying too little. The poet W. H. Auden pleaded to be told the truth about love:[1]

> Does it look like a pair of pyjamas,
> Or the ham in a temperance hotel?
> Does its odour remind one of llamas,
> Or has it a comforting smell?
> Is it prickly to touch as a hedge is,
> Or soft as eiderdown fluff?
> Is it sharp or quite smooth at the edges?
> O tell me the truth about love.

It is not sufficient to say that you will recognize love when you see it, rather like the helpful passer-by giving directions in the days before satnav ending with the assurance that 'You can't miss it'.

> When it comes, will it come without warning
> Just as I'm picking my nose?
> Will it knock on my door in the morning,
> Or tread in the bus on my toes?
> Will it come like a change in the weather?
> Will its greeting be courteous or rough?
> Will it alter my life altogether?
> O tell me the truth about love.

Humans need to be loved. In his great novel structured on the descendants of Adam and Eve, John Steinbeck expressed it poignantly. In expounding the Cain–Abel story on which the whole book is based, the Chinese intellectual servant Lee explains how it is the best-known story in the world because it is everybody's story, the symbol story of the human soul. 'The greatest terror a child can have is that he is not loved, and rejection is the hell he fears.'[2]

The ancient Greeks devised a fourfold classification of different kinds of love: (i) the love that derives from familiarity, including that within a family—such as the love of a parent for a child; (ii) deep friendship; (iii) romantic or erotic love; and (iv) unconditional love.[3] More recent classifications by psychologists and others have refined this, adding self-love, obsessive love, playful love (the early stages of romantic love), and—our favourite—enduring love (as exhibited by a couple who have been married for decades).

Throughout this book we have sought to discern what both scientific insight and spiritual wisdom can contribute to humanity's search for human flourishing. We have interpreted 'scientific insight' broadly. Within it, we have included not only lessons from the natural sciences but from the social sciences too. Scientific insight, broadly understood, has, as we have seen, much to tell us about the material and relational dimensions of human flourishing.

An evolutionary biologist might attribute all manifestations of love to the same root—namely improved ability to survive and reproduce.[4] Observing the faithfulness of swans (Figure 11.1),[5] the devotion of many animals (often the female) to their young, and the cooperation that one sees in social insects, prides of lions, and other species, it is easy to accept that human love has its origins in natural selection. Evolutionary psychologists generally observe that when the going gets tough, parental relationships are most robust with genetic children;[6] nevertheless, step-parents and the parents of adopted children can lavish love, time, care, and other resources on children who are not biologically theirs for many years. As with all other human qualities that we admire, as well as those that we deplore, we would expect the capacity for love to have come about through evolutionary processes, and to be shared in some measure with other species. But humans can rise above their genetic origins.[7] If any of us are to make any additional contribution to human flourishing, then it is up to us to choose how to use our capacity for loving. Anyone who has got this far in the book probably agrees, because

Figure 11.1 A number of animal species are more likely to be sexually faithful to one another than are many humans. Bewick's swans nearly always pair for life.

its purpose is to stimulate clear informed thinking about how to promote human flourishing.

We have also interpreted 'spiritual wisdom' broadly. There is a large literature on the relationship between religion and spirituality.[8] Religious believers tend to see spirituality as a core part of their religious belief and practice, but there are many who see themselves as spiritual without adhering to any formal religion. Although both authors are Christians, our intention in this book is not in writing apologetics for Christianity. Rather, we are more interested in showing how the transcendent dimension is a key part of human flourishing. One advantage of the term 'transcendent' is that it sits comfortably both within and without religion belief and practice. Indeed, there has been a growth in interest by atheists and others in attempting to discern what in religion might be of value without entailing a belief in God. Atheists such as Alain de Botton, in *Religion for Atheists: A Non-believer's Guide to the Uses of Religion*,[9] and at a more academic level Ronald Dworkin in *Religion without God*[10] are attempting to discern what in religion might be of value without entailing a belief in God.

From a theological perspective one would expect that if God is love, God would enable love to emerge in the story of life. Evolution is all about leaving copies of oneself in future generations—entities that do this come to predominate. While many species simply abandon (no criticism intended!) their offspring, others look after them. Such care is not far from love (the Latin *caritas* gives rise to both 'charity' and 'care'). An emotion like love (it might be more accurate to write 'emotions like the various types of love') could scarcely have arisen other than through evolution. It is no easier to know what goes on in a swan's mind than it is to know what it is like to be a bat,[11] but the two swans in Figure 11.1 may well feel something (perhaps a milder version) of what we feel when we fall in love. Many non-human species have parents who look after their young for long periods of time and, if these young go missing, behave not unlike human parents. Primatologist Frans de Waal and others have shown, perhaps unsurprisingly, that the data for animal emotions, including love, seem most convincing for our closest evolutionary relatives.[12]

The capacity for love, like the motivation for love, need not have a single origin. There can be more than one explanation for a given phenomenon. The answer to the question 'Why did my front door close one minute after I opened it?' depends on whether one is a physicist interested in Newtonian forces, a neuropsychologist interested in brain activity, or Michael observing that his 16-year-old cat, Hocus, takes rather longer nowadays to wander across the lawn and come in than he used to.

Complementary to evolutionary accounts of origins of what we recognize as love in humans are the accounts provided by the world's religions which, by and large, see God as the author of love. Christianity shares with Judaism the summary of the commandments 'Love the Lord your God with all your heart and with all your soul and with all your strength'[13] and 'love your neighbour as yourself.'[14] For all their differences, this latter call to unconditional love is found in some form in all the major world religions, including Buddhism, Christianity, Confucianism, Hinduism, Islam, Judaism, North American Spirituality, and Taoism.[15] It is expressed in the story of the good Samaritan, in the South African concept of *Ubuntu*, and in the refrain of Foy Vance's song 'Consider it an indiscriminate act of kindness'.

Christianity is emphatic that the source of love is God. As Charles Wesley put it in the first verse of a hymn written in 1747:

> Love divine, all loves excelling,
> Joy of heaven to earth come down,
> Fix in us thy humble dwelling,
> All thy faithful mercies crown.
> Jesu, thou art all compassion,
> Pure, unbounded love thou art;
> Visit us with thy salvation,
> Enter every trembling heart.

Cornerstone verses of the New Testament emphasize this: 'Dear friends, let us love one another, for love comes from God. Everyone who loves has been born of God and knows God. Whoever does not love does not know God, because God is love,'[16] and 'For God so loved the world that he gave his one and only Son, that whoever believes in him shall not perish but have eternal life.'[17] That is what Charles Wesley was writing about.

In the Christian tradition, the Word becoming physically embodied is both a realization of God's love and an inspiration for humans to love. Paul wrote to his friends in Philippi from prison in Ephesus or Rome, we don't know for sure which. Prison in those days was not a fixed-term punishment, but rather where people were held while those with power decided what to do with them. Paul quoted a song that was doing the rounds at the time to encourage his friends to think among themselves with the mind of the Messiah, Jesus, 'Who, though in God's form, did not regard his equality with God as something he ought to exploit.'[18] Instead, he let go of what he was entitled to, was born and grew up as a human, and eventual gave up his life through death on a cross. The Greek word here translated 'something he ought to exploit' has a connotation of clinging on to what you are entitled to for your own benefit. God's love is all about letting go for the benefit of others. It is about relinquishing our own entitlement to flourish in order to release resources to enable others to flourish, though sacrificial generosity has an astonishing habit of enriching the giver.

Human flourishing—empirical studies

Empirical studies on human flourishing, drawn from the social sciences, don't tell the whole story, but there is value in knowing what large numbers of individuals think and feel. An article by David Myers and Ed Diener addressing the question 'Who is happy?' has been cited over

3,500 times since it came out in 1995. The article reviewed what it described as a 'flood of new studies' that were exploring people's subjective well-being [19]

In 2018 Myers and Diener updated their earlier review.[20] Their central conclusions still stand. Although women typically experience substantially higher rates of anxiety and depression, and men typically engage in more antisocial conduct and misuse alcohol and drugs, women and men are remarkably similar in terms of how they evaluate either their happiness or their satisfaction with their life. A change since the 1960s is that the American dream has metamorphosed into life, liberty, and the *purchase* of happiness. When those entering US higher education are given a choice of nineteen possible life goals, 80 per cent rate as their first choice 'being very well-off financially', compared with less than 45 per cent in the 1960s (Figure 11.2).

The polymath Tyler VanderWeele has degrees in mathematics, philosophy and theology, finance and applied economics, and biostatistics. He includes measures of happiness in his work but tries to go beyond them to explore more fully what contributes to human flourishing— still through the lens of empirical studies in the social sciences.

Figure 11.2 Percentage of new college students in the US identifying two of nineteen possible life goals.

VanderWeele points out that individual disciplines like medicine, public health, psychology, and economics typically focus on narrow outcomes—recovery from a particular disease, happiness, the consumption of goods and services, or whatever.[21] By contrast, human flourishing relates to life taken as a whole.

How people answer questions about how well they are doing depends on the context. The philosopher Anna Alexandrova imagines 24 hours in the life of a heavily pregnant woman called Masha.[22] Masha slips on an icy street and hurts her knee. A Good Samaritan rushes to help her up and takes her to a bench to recover. He asks, 'How are you?' She replies that she will be OK when she has rested. That evening in a quiet moment at a dinner party a longstanding friend asks Masha, 'How are you?' Masha tells her all about her abandoned PhD, the challenges in her marriage, and the likely impact of the baby on her career. The next day a prenatal social worker visits and asks, 'How are you?' Masha talks about family support for the early stages of caring for the baby after it is born. Each of these three is asking about her well-being, but the meaning of the question depends on who they are and the context in which they are asking it. It also depends on the values against which well-being is measured.

Self-reported broad measures of psychological well-being or life satisfaction are subjective measures as described by the individuals themselves—after all, if you say that you are satisfied with your life, who am I to disagree? And yet, we might ask, does not the notion of human flourishing suggest an interplay between subjectivity and objectivity?

> a person may feel satisfied with life, and yet be utterly depraved, or without meaningful social relationships, or entirely dependent upon narcotics. Would we say such persons are flourishing?[23]

This suggests that while flourishing involves emotions it goes beyond them. Flourishing is a state in which all aspects of a person's life not only *feel* good but *are* good. VanderWeele offers measurements based on five broad domains of human life: (i) happiness and life satisfaction; (ii) health, both mental and physical; (iii) meaning and purpose; (iv) character and virtue; and (v) close social relationships; plus a sixth denoting the financial and material stability necessary to sustain the first five. Unsurprisingly, these overlap both with the three dimensions to human flourishing that we proposed in Part I and with the three underpinning concepts of truth, purpose, and meaning that we identified in Part II. Who is best placed to assess how these domains are going? Humans have a great capacity both for ignorance and for self-deception. If someone is delusional, their own views about their mental health are, almost by definition, not ones in which someone else should have a great deal of confidence as to their veracity. If I am really interested in another person's mental and physical health, I might value the opinion of a professional as well as that person's views. Self-assessment has its uses, but its components should be open to more objective measurement.

VanderWeele identifies four pathways to human flourishing: family, work, education, and religious community. Two of these elements have been attributed to Sigmund Freud: '*zu lieben und zu arbeiten*' ('to love and to work').[24] In recognizing the contributions that these pathways can make to human flourishing, their intrinsic value should not be lost. However firmly the pursuit of happiness is enshrined in the United States Declaration of Independence, it remains an elusive goal. Those who seek it may all too easily find that they lose it. Happiness generally comes to those who work towards something of enduring value.

Anyone who decides to spend more time with their family *in order* to flourish is rather missing the point. Similarly, if you participate in communal worship more frequently with the expectation that you will become less depressed, live longer, or be less likely to develop cancer, your expectations may be fulfilled but that's not the purpose of worship. The deeper reasons for worship are more to do with the relational and transcendent dimensions of flourishing.

Love in action

One of the defining moments of the 1968 Mexico Olympic Games came when two African-American athletes, Tommie Smith and John Carlos, each raised a black-gloved fist during the playing of the US National Anthem in the medal ceremony that followed the men's 200 m. Smith had just won, in a world-record time of 19.83 s, and Carlos had come third. At a press conference after the event Tommie Smith said: 'If I win I am an American, not a black American. But if I did something bad then they would say "a Negro". We are black and we are proud of being black.'[25] Smith and Carlos were booed by the crowd as they left the podium and expelled from the Games on the grounds that their behaviour was a deliberate and violent breach of the fundamental principles of the Olympic spirit. Although the African-American community hailed them as heroes, they were largely ostracized by the US sporting establishment and received death threats.

The Australian Peter Norman finished second in the race (Figure 11.3). Norman knew that Smith and Carlos were going to protest at the medal ceremony and asked a member of the U.S. rowing team for his 'Olympic Project for Human Rights' badge, so that he could show solidarity. Norman returned to Australia a pariah and suffered unofficial sanction.[26] When Norman died, in 2006, the two pallbearers at the front carrying his coffin were Tommie Smith and John Carlos. In his eulogies, Carlos recounted the conversation the three of them had before going out for the medal ceremony. They asked Norman if he believed in

Figure 11.3 Peter Norman, the Australian silver medal winner in the men's 200 m, supported Tommie Smith and John Carlos in their Olympic Project for Human Rights protest. His principled actions, a manifestation of disinterested love, had substantial adverse repercussions for him.

human rights. He said he did. They asked him if he believed in God. Norman, who came from a Salvation Army background, said he believed strongly in God. Carlos related how they knew that what they were going to do was far greater than any athletic feat. Norman affirmed, 'I'll stand with you.' Carlos said he expected to see fear in Norman's eyes. He didn't. 'I saw love.'[27] All this was more than half a century before such issues were articulated as 'Black Lives Matter'.

Much love, just like much flourishing, need not be disinterested; usually there are benefits to all parties. This is most evident in a close friendship or in a marriage or the equivalent. When we first meet, you like gardening and I have no particular interest in gardening, whereas I like Outsider Art and you have barely heard of it. Half a lifetime later, I am doing much of the gardening, and enjoying it, and we jointly collect Outsider Art. But what Norman did seems a clear instance of disinterested love—it was perfectly evident to him that he was not going to benefit from his actions; his motivation was that others should.

Effective altruism

Effective altruism is a movement in which people aspire to use evidence and reason to do as much good as possible.[28] It has more in common with Peter Norman's principled action than with a love between two people with mutual benefit. It begins with a question: How can we use our resources (time and money) to help others the most? To see the direction of travel of the answer, consider the following:

> Imagine if, one day, you see a burning building with a small child inside. You run into the blaze, pick up the child, and carry them to safety. You would be a hero. Now imagine that this happened to you every two years—you'd save dozens of lives over the course of your career.
>
> This sounds like an odd world. But current evidence suggests it is the world that many people live in. If you earn the typical income in the US, and donate 10% of your earnings each year to the Against Malaria Foundation, you will probably save dozens of lives over your lifetime.[29]

This is about analysis and the exercise of willpower. The question is then, for effective altruists, which charities to support. One approach is to choose a charity that maximizes the number of additional quality-adjusted life years (QALYs) for one's donation. QALYs are widely used in evaluating the benefit of an intervention in health care. Most people want not simply to live longer but rather to have more years of good health, assessed in terms of the ability to carry out the activities of daily life, and freedom from pain and mental disturbance.

Effective altruists don't only believe in helping humans. They point out that those who work to reduce animal suffering can benefit huge numbers of sentient individuals. For example, the work that Compassion in World Farming[30] has done—and even advisory committees like the Farm Animal Welfare Council/Committee[31] in the UK—benefits literally billions of animals each year.

The strength and the weakness of effective altruism is that it is rational. In the absence of rationality it is all too easy for any of us to persuade ourselves that we don't have quite enough spare time or money to do more than occasionally give a very small amount of either to the latest one-off appeal for an emergency fund or other cause. But altruism also benefits from the affective dimension and the ties of relationships. There is something rather anonymous about effective altruism. It is hard to image effective altruists condoning any of us spending large amounts of money on those who are close to us, given the perceived need to alleviate world poverty or fight against threats to the very survival of humanity.

The approach of effective altruism looks like a kind of utilitarianism in which the utility (the goal to be maximized) is the sum of QALYs over the world population. The utility could be further increased by increasing the world population until individual QALYs became so low that the sum started to decrease. But then one would need to think about future generations, and perhaps apply a discount rate to their QALYs. A rigorous approach to population ethics invokes the concept of *Total Utilitarianism*, which invites members of each generation to evaluate what the future world population should be on the basis of a weighted sum of well-being across the generations.[32] The discount rate to be applied in the weighting need not imply that future generations have a lower value; it could simply mean that the uncertainty of the effect of actions is greater the further ahead one tries to calculate.

There is even a formula, for those who enjoy such things (the rest of us can skip to the next paragraph). Social well-being is calculated as the sum of each person's well-being from the time of evaluation t to infinity. If $N(t)$ is the number of adults at t, and the number of children born by then is $N(t+1)$, and they all have the same well-being as a function of consumption $U(C)$, with a discount rate Θ (which could be a discount for the risk of extinction), then the total well-being across the generations is[33]

$$V^{U}(t) = \sum_{u=t}^{\infty} [(N(u)) + (N(u+1))U(C(u))]\Theta^{(u-t)}.$$

Depending on the assumptions, Total Utilitarianism has been calculated to commend an optimal population at any one time anywhere in the range 5.9 billion to 142 billion.[34] Maybe further considerations should come into play, such as space for wildlife.

Other criteria lead to different results for a sustainable population based on the total economic capacity of the Earth.[35] On that basis, for an average income of 20,000 international dollars per person, it has been calculated that the population would need to be limited to 3.5 billion, well below the current population, though greater than it was in 1960. If an average income of 10,000 international dollars were acceptable, which is below the current average, then the sustainable population is calculated to double to 7 billion, a bit less than it is now, and substantially less than the predicted population at the turn of the next century. By considering only averages, we neglect the devastating effect on those at the bottom if the variance in living standards continues to grow.[36] Considerations of global climate change and sustainable energy consumption introduce further constraints. We are not too worried by the validity of the precise numbers predicted by this approach. Just trying to find the criteria for the optimum size of the global population concentrates the mind on fundamental questions about what it really means for humans to flourish.

Another way to bring the questions into focus is through a thought experiment inspired by P. D. James' novel *The Children of Men*, which we introduced in Chapter 6. Imagine that you learned that although you would live a normal lifespan, one month after you died humanity would be wiped out in a collision with a giant asteroid. Leaving aside questions such as how you know, and whether anyone else knows, and assuming that no intervention could prevent this fate, how would this certain knowledge affect your attitudes during the remainder of your life? Which activities would no longer seem worthwhile? Which would you wish to modify, and how? And what new activities might you choose to undertake? Samuel Scheffler, who devised this morbid thought experiment, does not believe in an afterlife as understood by many religious people; he believes rather that biological death represents the final and irrevocable end of an individual's life. Nevertheless, he is convinced that the continuity of life—other people's lives—after our own death matters greatly to us.

The prospect is unlikely to engender indifference. It might also seem fruitless to try to evaluate whether the end of suffering would more than offset the end of joy. A more likely reaction is an overwhelming sense of dismay at the prospect that the people (and other entities) we care about would cease to exist, or even have the possibility to flourish. And we would almost certainly re-evaluate which of the projects into which we now pour so much of ourselves would still seem worthwhile. Scheffler reckons that the scenario would lead to 'a world characterized by widespread apathy, anomie, and despair; by the erosion of social institutions and social solidarity; by the deterioration of the physical environment; and by a pervasive loss of conviction about the value or point of many activities.'[37]

Happily, this was only a thought experiment. It remains possible that human extinction may happen, conceivably within our lifetimes—or rather, by definition, at the end of them—whether by error, terror, or some other cause beyond human control. That is why it is important to consider existential threats,[38] and, more positively, to consider long-term effects in policy-making.[39] The extinction of humanity would be a disaster of a different kind from the death of even very large numbers of people. For half the world population to die in a pandemic would be terrible, but, in a way that it is hard to put one's finger on, for the human race to become extinct would seem more than simply twice as bad, and more than 1 per cent worse than 99 per cent of the world's population dying. We somehow accept that we and everyone we love and care about will die; that billions of people we don't or shan't know might die or never live provokes a different reaction. The thought experiment has the power to reveal that there is in most of us a strong desire that humans should flourish, and should continue to flourish.

Existential threats to human flourishing

We live in a time where humanity is particularly threatened.[40] In 1947 the scientists who had worked to develop the first atomic weapons in the Manhattan Project created the Doomsday Clock, using the imagery of apocalypse (equated with midnight on the clock) to convey threats to humanity and the planet. Back then, they set the clock at seven minutes to midnight (i.e., 2353). The decision as to whether to move the minute hand of the Doomsday Clock is made every year by the Bulletin of the Atomic Scientists' Science and Security Board in consultation with its Board of Sponsors, which includes 13 Nobel laureates. On 23 January 2020, the time was set at 100 seconds to midnight, the closest it has ever been to apocalypse and the first time it had been set other than in whole minutes (Figure 11.4). As the first paragraph of the official 2020 Domesday Clock statement put it:

> Humanity continues to face two simultaneous existential dangers—nuclear war and climate change—that are compounded by a threat multiplier, cyber-enabled information warfare, that undercuts society's ability to respond. The international security situation is dire, not just because these threats exist, but because world leaders have allowed the international political infrastructure for managing them to erode.[41]

An existential threat is understood as one that has the capacity to wipe out all (or most) of humanity. Long a focus of science-fiction writers (think H. G. Wells' 1897 *The War of the Worlds* onwards), existential threats are increasingly the focus of academic research. The Centre for the Study of Existential Risk is an interdisciplinary research centre at the University of Cambridge, one of an increasing number of centres looking at such risks.[42] Its work includes topics we chose for Chapter 10: risks from artificial intelligence, risks from misuse of genetic engineering, and risks from natural pandemics. Other risks under study include catastrophic ecosystem collapses, runaway climate change, and risks to global justice driven by inequality, corruption, and structural discrimination.[43]

There seem to be two main conclusions that can be drawn from research on existential threats. First, the lessons of the past, quite apart from the predictions of the future, strongly suggest that disasters are going to occur that are beyond what most of us in countries that are not presently ravaged by war or starvation tend to imagine. Furthermore, we are talking about a time scale of less than a century—some of us will not be around, but the next generation and the one after that will be. Secondly, *Homo sapiens* will probably not be wiped out—so think of the Earth after the flood in *The Epic of Gilgamesh*, Genesis, Sura 71 of the Quran, and other scriptures, or the post-apocalyptic worlds beloved by science-fiction authors (from Mary Shelley's *The Last Man* onwards) and film-makers (*Planet of the Apes*, the Mad Max series, and *Interstellar*) rather than extinction.

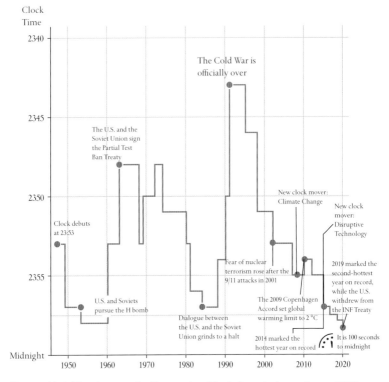

Figure 11.4 The time on the Doomsday Clock from its inception to 2020.

Homelessness

The last global survey of homelessness was undertaken in 2005—embarrassingly long ago. It was estimated at that time that approximately 100 million people worldwide were without a place to live, and over a billion people were inadequately housed.[44] Just as thought experiments about human extinction, and academic studies of its actual risk, can focus the mind on what really matters to us, so can an issue which is rather closer to home (pun intended) for many people than predictions about nuclear annihilation, runaway climate change, asteroid impacts, pandemics, or whatever future disaster might affect humanity (Figure 11.5). Precisely what is meant by 'homelessness' varies between countries, but whatever the context, as we mentioned in Chapter 2, it is hard to assert that a homeless person is flourishing as they could be.

Figure 11.5 There are various types of homelessness, with rough sleeping the most visible. Homelessness can be tackled though systematic care.

The United Nations defines a homeless household as 'those households without a shelter that would fall within the scope of living quarters. They carry their few possessions with them, sleeping in the streets, in doorways or on piers, or in another space, on a more or less random basis.'[45] Some homeless people are, literally, on the streets; others rely on friends, family or emergency accommodation without any expectation that such arrangements have any degree of permanence.

The causes of homelessness say as much about wider society as about those who are homeless. Longitudinal studies show that the risks of homelessness are often rooted in an individual's childhood and are not infrequently to do with violence or other types of abuse in the parental home. These roots are often fertilized by an unwholesome diet of subsequent problems to do with alcohol, other drugs, a shortage of money, and the breakdown of relationships.[46]

Homelessness can be of short duration. If it lasts for more than a matter of a few weeks it becomes increasingly difficult to escape from. In many Western countries, being homeless makes it difficult to gain steady employment, to receive more than emergency medical care, and to have access to a bank or other method of organizing one's finances.

Unsurprisingly, homelessness is associated with poorer physical and mental health, drug misuse, lower measures of self-worth, and a loss or shift in identity.

A home is not simply a physical space. For most of us, our home is part of our identity, somewhere to where we return as a place of safety and more than that—of centredness. This is as true for nomadic people, like the Bedouin, who traditionally take their home with them as they move around their environment, as it is for the majority of people whose homes are geographically fixed. For many people, their home is where their closest relationships take place, with a partner, children, or extended family.

Given that having a home is good for body, mind, and spirit, is homelessness something we have to accept? Even in wealthy countries free from war and such natural disasters as widespread earthquakes or flooding, homelessness seems endemic. In most countries, the problem is getting worse—Finland is a welcome exception, where systematic efforts have been made since the 1980 to tackle the problem.[47] Various plans by charities and other organizations exist to prevent or end homelessness, including a raft of interventions, such as more affordable housing, legal protection for homeless people, and the provision of short-term emergency accommodation.[48]

The advent of COVID-19 led some governments rapidly to prioritize action for the homeless, as it was appreciated that homeless people were likely to be particularly vulnerable to infection. In the UK, within weeks there was a major response including the provision of hotel and emergency accommodation, the suspension of evictions from Home Office asylum accommodation, and the halting of evictions from the private and social rented sectors.[49] This led to very substantial reductions in homelessness, with about a half of all rough sleepers offered accommodation within weeks.[50] It has been argued that a universal basic income would virtually eliminate homelessness and deal with a raft of other social problems while being relatively affordable and a socially just solution.[51]

The material and relational dimensions of homelessness are clear enough—by and large both need to have gone awry for someone to end up homeless for any period of time. But is there a transcendental aspect too? The New Testament describes the world to come as a better country: 'Instead, they were longing for a better country—a heavenly one. Therefore, God is not ashamed to be called their God, for he has prepared a city for them.'[52] The idea here is that deep down, however wonderful this world and our own homes on it may be, there may, perhaps should, remain a sense of dissatisfaction, of alienation. In the Jewish scriptures, David, when he seemed settled in Jerusalem and preparations were being made to build the first temple, is recorded as saying, 'We are foreigners and strangers in your sight, as were all our ancestors. Our days on earth are like a shadow, without hope.'[53]

That the author of Hebrews talks of the world to come as being a city may surprise those of us who are more likely to associate contemporary cities with noise and pollution and the countryside with escape and quietness. When people are asked to describe their favourite paintings, traditional rural landscapes come out top.[54] An international research project, 'The People's Choice', undertaken by two Russian artists, Vitaly Komar and Alexander Melamid, identifies open landscapes with lakeside scenes as the most popular[55]—something which evolutionary psychologists, who get everywhere, have claimed is connected to our evolutionary history. But the book of Revelation also pictures the world to come as a city, the New Jerusalem, Zion. Cities are often seen as beacons of hope. The history of civilization—as the word suggests— is largely coterminous with cities, which only really came into existence some 5000 years ago. In 2018 about 55 per cent of the world's population lived in cities; by 2100, the figure is projected to be 85 per cent.[56]

In 410, Rome, known by then as 'The Eternal City', was captured by the Vandals, presaging the fall of the Roman Empire over the new few decades. Three years later Augustine (354–430), Bishop of Hippo from 396 till his death, began his book *The City of God*. The first half of *The City of God* defends Christians against the charge that they were responsible for the fall of Rome. In the second half, Augustine contrasts an earthly city, like Rome, with a heavenly one. The heavenly city is seen as our true destination. Earlier, in his *Confessions*, Augustine had written 'You have made us for yourself, O Lord, and our heart is restless until it finds its rest in you.' The Lord's Prayer asks that the Kingdom of Heaven should come to Earth. In *The City of God* Augustine argues that the peace and happiness to be found in the heavenly city can also be experienced here on Earth. There is a continuity between this world and the next.

Love as the essence of flourishing

How does love provide not only a pattern of human flourishing but a resource for human flourishing? If love were fuel, like petrol or hydrogen, how could we imagine it powering the engine of human flourishing? Without getting carried away by the metaphor, can love be a renewable resource for flourishing?

We believe that it can. We also find that how to answer that question goes to the very depths of who humans are and what foundational beliefs our lives rest on. The two of us think that we can therefore most usefully address the question by being true to ourselves, and hope that readers will find the worked exercise useful in being true to themselves.

In the Judeo-Christian tradition, two instructions about love float to the top: love your God, and love your neighbour. There are warnings about too broad a view of God, and too narrow a view of neighbour. Each demands scientific insight and spiritual wisdom.

Time and again Jewish thought leaders emphasized that love for God was to be manifested in practical action. It meant not accumulating excess material goods—in the eighth century BCE ivory furniture featured prominently in desirable luxuries for the rich[57]—but rather attending to the material needs of others. That requires a combination of scientific insight, including what we would now call the social sciences, and spiritual wisdom—why should we care for those who cannot help themselves? But it goes further.

In 2019 Greta Thunberg spoke to world leaders at the United Nations Climate Action Summit in New York:

> This is all wrong. I shouldn't be up here. I should be back in school on the other side of the ocean. Yet you all come to us young people for hope. How dare you!
>
> You have stolen my dreams and my childhood with your empty words. And yet I'm one of the lucky ones. People are suffering. People are dying. Entire ecosystems are collapsing. We are in the beginning of a mass extinction, and all you can talk about is money and fairy tales of eternal economic growth. How dare you![58]

Many people care about climate change because of the direct impact on human flourishing. According to the 2018 Intergovernmental Panel on Climate Change (IPCC) *Special Report on Global Warming of 1.5°C*, even with that temperature rise above historical levels there would be increased risks and likely damage to health, livelihoods, food security, water supply, human security, and economic growth.[59] These all matter because they affect huge numbers of lives, among them some of the most vulnerable in the world. Further predictions indicate that a 2°C temperature increase would exacerbate extreme weather, rising sea levels and diminishing Arctic sea ice, coral bleaching, and loss of ecosystems, among other impacts. Why are these important? Why does coral bleaching matter? In part it is because of the impact on tourist industries—tour guides for the Australian Barrier Reef may lose their livelihoods, and because of diminished aesthetic pleasure—scuba divers will no longer enjoy the wonderful colours of living coral. Other industries, including fishing, will suffer. These are human-centred considerations, and they have weight.

There is another basis for caring if ecosystems are considered valuable in themselves. We might then care about extinctions of species because we believe that the species themselves have value, and that their loss is somehow deeply significant. For those who speak of the Creation, this may be tied in with accountability to the Creator. This was what motivated Sir John Houghton, one of Britain's most distinguished climate scientists, who died in 2020. He wrote that 'although my science and my faith always sat quite comfortably together, it wasn't until I started working more specifically on climate change that they began to interweave.'[60] Like the IPCC of which he was a leading member, he was clear about the evidence and humble about the uncertainties. Spiritual wisdom gives a mandate for caring for the Creation; scientific insight provides the tools with which to discharge that mandate.

If scientific insight and spiritual wisdom came together in Sir John Houghton's passion for the climate, how much more should they support our passion for people! Love, so far from being a wishy-washy concept that can mean whatever you want it to, provides the fundamental driver for caring about human flourishing. The curious child may ask, 'Why does God love me?' The wise parent can answer, 'I don't know; God just does.' Scientific insight and spiritual wisdom can help us to know how we should promote human flourishing and why we should promote human flourishing. But they cannot make us decide to promote human flourishing. That is a choice for all of us, individually and as a society. If we choose well, it will be because of love.

Notes

1 Auden, W. H. (1999) *Tell Me the Truth about Love*, London: Faber.

2 Steinbeck, J. (1952/1963) *East of Eden*, London: Pan Books, London, p. 257.

3 One of the best-known exponents is Lewis, C. S. (1960) *The Four Loves*, London: Geoffrey Bles. The cultural context for some of what he writes has changed, but many of his insights are timeless.

4 For a standard introduction to evolutionary psychology, see Buss, D. (2019) *Evolutionary Psychology: The New Science of the Mind*, 6th edn, New York: Routledge, New York. For a recent critique of evolutionary psychology, see Smith, S. E. (2020) Is evolutionary psychology possible?, *Biological Theory* **15**, 39–49.

5 Evans, M. E. (1979) Aspects of the life cycle of the Bewick's Swan, based on recognition of individuals at a wintering site, *Bird Study* **26**(3), 149–62.

6 Daly, M. and Wilson, M. (1996) Evolutionary psychology and marital conflict: The relevance of stepchildren, in D. M. Buss and N. Malamuth (Eds) *Sex, Power, Conflict: Feminist and Evolutionary Perspectives*, New York: Oxford University Press, pp. 9–28.

7 Alexander, D. R. (2020) *Are We Slaves to our Genes?*, Cambridge, UK: Cambridge University Press.

8 See, for example, Pargament, K. I., Mahoney, A. Exline, J. J., Jones, J. W. and Shafranske, E. P. (2013) Envisioning an integrative paradigm for the psychology of religion and spirituality, in K. I. Pargament, J. J. Exline, and J. W. Jones (Eds), *APA Handbook of Psychology, Religion, and Spirituality*, Vol. 1: *Context, Theory, and Research*, American Psychological Association, Washington, DC, pp. 3–19.

9 De Botton, A. (2012) *Religion for Atheists: A Non-believer's Guide to the Uses of Religion*, London: Penguin.

10 Dworkin, R. (2013) *Religion without God*, Cambridge, MA: Harvard University Press.

11 Nagel, T. (1974) What is it like to be a bat?, *The Philosophical Review* **83**, 435–50.

12 De Waal, F. (2019) *Mama's Last Hug: Animal Emotions and What They Tell Us about Ourselves*, New York: W. W. Norton.

13 Deuteronomy 6: 5.

14 Leviticus 19: 18.

15 Templeton, J. M. (1999) *Agape Love: A Tradition Found in Eight World Religions*, West Conshohocken, PA: Templeton Press.

16 1 John 4: 7–8.

17 John 3: 16.

18 Philippians 2: 5–11.

19 Myers, D. G. and Dieners, E. (1995) Who is happy?, *Psychological Science* **6**(1), 10–19, p. 10.

20 Myers, D. G. & Dieners, E. (2018) The scientific pursuit of happiness, *Perspectives on Psychological Science* **13**, 218–25.

21 VanderWeele, T. J. (2017) On the promotion of human flourishing, *Proceedings of the National Academy of Sciences* **114**(31), 8148–56.

22 Alexandrova, A (2017) *A Philosophy for the Science of Well-Being*, Oxford: Oxford University Press, pp. 6–8.

23 VanderWeele (2017) On the promotion of human flourishing, p. 8149.

24 It has been suggested that Erik Erickson is responsible for the epigram as it was he who first recounted it a decade after Freud had died. See Bill (2012) Lieben Und Arbeiten, *Figuring Out Fulfilment: A Storytelling Project about Finding a Career and Finding Yourself*. 23 January. Available at https://figuringoutfulfillment. wordpress.com/2012/01/23/lieben-und-arbeiten/.

25 BBC (n.d.) 1968: Black athletes make silent protest, *On This Day: 17 October*. Available at http://news.bbc.co.uk/onthisday/hi/dates/stories/october/17/ newsid_3535000/3535348.stm.

26 Montague, J. (2012) The third man: The forgotten Black Power hero, *CNN*. Available at https://edition.cnn.com/2012/04/24/sport/olympics-norman-black-power/index.html.

27 Flanagan, M. (2006) 'Tell your kids about Peter Norman', *The Age*, 10 October. Available at https://www.theage.com.au/national/tell-your-kids-about-peter-norman-20061010-ge3axk.html.

28 Ord, T. (2020) *The Precipice: Existential Risk and the Future of Humanity*, London: Bloomsbury, p. 8.

29 Effective Altruism (2016) Introduction to effective altruism. Available at https://www.effectivealtruism.org/articles/introduction-to-effective-altruism/.

30 Compassion in World Farming, https://www.ciwf.org.uk.

31 Farm Animal Welfare Committee, https://www.gov.uk/government/groups/ farm-animal-welfare-committee-fawc—now the Animal Welfare Committee (with an expanded brief), https://www.gov.uk/government/groups/animal-welfare-committee-awc.

32 Dasgupta, P. (2019) *Time and the Generations: Population Ethics for a Diminishing Planet*, New York: Columbia University Press, p. 78.

33 Ibid., p. 90, equation 31.

34 Ibid., pp. 111–12.

35 Dasgupta, A. and Dasgupta, P. (2019) Socially embedded preferences, environmental externalities, and reproductive rights, in *Time and the Generations: Population Ethics for a Diminishing Planet*, New York: Columbia University Press, pp. 258–9.

36 Collier, P. (2008) *The Bottom Billion: Why the Poorest Countries are Failing and What Can Be Done About It*, Oxford: Oxford University Press.

37 Scheffler, S. (2013) The afterlife (Part I), in S. Scheffler and N. Kolodny, *Death and the Afterlife*, Oxford: Oxford University Press, p. 18.

38 The Centre for the Study of Existential Risk, https://www.cser.ac.uk/; Szocik, K., Norman, Z. and Reiss, M. J. (2020) Ethical challenges in human space missions: A space refuge, scientific value, and human gene editing for space, *Science and Engineering Ethics* **26**, 1209–27.

39 All-Party Parliamentary Group for Future Generations, https://www.appg-futuregenerations.com/.

40 Rees, M. (2004) *Our Final Century*, New York: Penguin Random House; Rees, M. (2018) *On the Future: Prospects for Humanity*, Princeton, NJ: Princeton University Press.

41 Science and Security Board Bulletin of the Atomic Scientists (2020) *Closer than Ever: It is 100 Seconds to Midnight—2020 Doomsday Clock Statement*. Available at https://thebulletin.org/doomsday-clock/2020-doomsday-clock-statement/.

42 See also the Future of Humanity Institute, https://www.fhi.ox.ac.uk, and the Future of Life Institute, https://futureoflife.org/background/existential-risk/?cn-reloaded=1.

43 Centre for the Study of Existential Risk, https://www.cser.ac.uk/research/.

44 United Nations Economic and Social Council (2005) *Economic, Social and Cultural Rights: Report of the Special Rapporteur on Adequate Housing as a Component of the Right to an Adequate Standard of Living*. Available at https://documents-dds-ny.un.org/doc/UNDOC/GEN/G05/117/55/PDF/G0511755.pdf?OpenElement.

45 United Nations, Department of Economic and Social Affairs Statistics Division, Demographic and Social Statistics Branch (2004) *United Nations Demographic Yearbook Review: National Reporting of Household Characteristics, Living Arrangements and Homeless Households—Implications for International Recommendations*. Available at https://unstats.un.org/unsd/demographic/products/dyb/techreport/hhChar.pdf.

46 Ravenhill, M. (2008) *The Culture of Homelessness*, Farnham, UK: Ashgate.

47 De Oliveira, B. (2017) *Universal Basic Income: A Policy to Reduce Homelessness in the UK?* Available at https://medium.com/@brunodeoliveira_14513/universal-basic-income-a-policy-to-reduce-homelessness-in-the-uk-bd9d4b6f9081.

48 Downie, M., Gousy, H., Basran, J., Jacob, R., Rowe, S., Hancock, C. et al. (2018) *Everybody In: How to End Homelessness in Great Britain*, London: Crisis.

49 Crisis (2020) Government response to homelessness and COVID-19. Available at https://www.crisis.org.uk/media/241941/crisis_covid-19_briefing_2020.pdf.

50 Wall, T. (2020) How a hotel is stemming the tide of Covid-19 among rough sleepers, *The Guardian*, 5 May. Available at https://www.theguardian.com/society/2020/may/05/covid-19-hotel-homeless-people-uk.

51 Wilderquist, K. (2019) End the threat of economic destitution now, Basic Income Earth Network, 22 October. Available at https://basicincome.org/news/2019/10/end-the-threat-of-economic-destitution-now/.

52 Hebrews 10: 16.

53 1 Chronicles 29: 15.

54 ArtChain Global (2018) Top 10 most popular subjects in paintings. Available at https://medium.com/@artchain_global/top-10-most-popular-subjects-in-paintings-1797e91a0d4a.

55 Sutton, B. (2018) This is America's most wanted painting, Artsy, 5 November. Available at https://www.artsy.net/article/artsy-editorial-komar-melamid-americans-painting-thought-wanted. See also https://awp.diaart.org/km/index.html.

56 European Commission (n.d.) Knowledge for policy: Urbanisation world-wide. Available at https://ec.europa.eu/knowledge4policy/foresight/topic/continuing-urbanisation/urbanisation-worldwide_en.

57 Amos 6: 4.

58 Thunberg, G. (2019) *No One Is Too Small to Make a Difference*, London: Penguin. Text of speech available at https://www.npr.org/2019/09/23/763452863/transcript-greta-thunbergs-speech-at-the-u-n-climate-action-summit?t=1592737516581.

59 Masson-Delmotte, V. (2018) *Global Warming of 1.5°C: An IPCC Special Report on the Impacts of Global Warming of 1.5°C above Pre-industrial Levels and Related Global Greenhouse Gas Emission Pathways, in the Context of Strengthening the Global Response to the Threat of Climate Change, Sustainable Development, and Efforts to Eradicate Poverty.* Geneva: Intergovernmental Panel on Climate Change.

60 Houghton, J. and Taylor, J. (2013) *In the Eye of the Storm: The Autobiography of Sir John Houghton*, Oxford: Lion Hudson, p. 223; quoted by White, R. (2020) Obituary: Sir John Houghton FRS (1931–2020), *Science & Christian Belief* **32**, 188–92.

Picture Credits and Sources

Index

Note: Figures are indicated by an italic *f* following the page number.